T0270887

STATISTICAL INFERENCE
IN STOCHASTIC PROCESSES

PROBABILITY: PURE AND APPLIED

A Series of Textbooks and Reference Books

Editor

MARCEL F. NEUTS

University of Arizona
Tucson, Arizona

STATISTICAL INFERENCE
IN STOCHASTIC PROCESSES

EDITED BY

N. U. PRABHU
**Cornell University
Ithaca, New York**

I. V. BASAWA
**University of Georgia
Athens, Georgia**

Marcel Dekker, Inc.   **New York · Basel · Hong Kong**

Library of Congress Cataloging--in--Publication Data

Statistical inference in stochastic processes / edited by N. U. Prabhu,
   I. V. Basawa.
         p.   cm. -- -- (Probability, pure and applied; 6)
      Includes bibliographical references and index.
      ISBN 0-8247-8417-0
      1. Stochastic processes. 2. Mathematical statistics. I. Prabhu,
   N. U. (Narahari Umanath).   II. Basawa, Ishwar V.
   III. Series.
   QA274.3.S73   1990
   519.2-- --dc20   90--20939
                                                              CIP

This book is printed on acid-free paper.

MARCEL DEKKER, INC.
270 Madison Avenue, New York, New York 10016

Current printing (last digit):
10 9 8 7 6 5 4 3 2 1

PRINTED IN THE UNITED STATES OF AMERICA

Preface

Currently, research papers in the area of statistical inference from stochastic processes ("stochastic inference," for short) are published in a variety of international journals such as *Annals of Statistics*, *Biometrika*, *Journal of the American Statistical Association*, *Journal of the Royal Statistical Society B*, *Sankhya A*, *Stochastic Processes and Their Applications*, and *Theory of Probability and Its Applications*. Some important papers have also appeared with limited visibility in national journals such as *Australian Statistical Association Journal*, *Journal of the Institute of Statistical Mathematics* (*Japan*), and the *Scandinavian Statistical Journal*. These journals are mostly interested in papers that emphasize statistical or probability modeling aspects. They seem to view stochastic inference as peripheral to their domains, and consequently provide only limited scope for publication of ongoing research in this important area.

The area of stochastic inference is an interface between probability modeling and statistical analysis. An important special topic within this broad area is time series analysis, where, historically, modeling and analysis have gone hand in hand; this topic is now well established as an area of research, with an identity of its own with periodic conferences, publication of proceedings and its own international journal. A second topic that has been successful in this regard is econometrics. However, this balanced development of modeling and inference has not taken place in other areas such as population growth, epidemiology, ecology, genetics, the social sciences, and operations research.

The great advances that have taken place in recent years in stochastic processes such as Markov processes, point processes, martingales, and spatial processes have provided new, powerful techniques for extending the classical inference techniques. It is thus possible to open up lines of communication between probabilists and statisticians.

Researchers in the area of stochastic inference keenly feel the need for some kind of distinct visibility. This fact is evident from recent national and international conferences at which separate sessions were organized on this topic. A "mini-conference" on stochastic inference was held at the University of Kentucky, Lexington, during Spring 1984. A major international conference on inference from stochastic processes was held in Summer 1987 at Cornell University, Ithaca, New York.

Although the need for distinct visibility for research on stochastic inference is strong enough, the market conditions do not fully justify the starting of a new journal on this topic at the present time. Rather, we felt that a series of occasional volumes on current research would meet the current needs in this area and also provide an opportunity for speedy publication of quality research on a continuing basis. In this publication program, we plan to invite established authors to submit survey or research papers on their areas of interest within the broad domain of inference from stochastic processes. We also plan to encourage younger researchers to submit papers on their current research. All papers will be refereed.

The present volume is the first in this series, under our joint editorship. (The proceedings of the Cornell Conference, edited by Prabhu, have appeared as *Contemporary Mathematics*, Volume 80, published by the American Mathematical Society.) We take this opportunity to thank the authors for responding to our invitation to contribute to this volume and gratefully acknowledge the referees' services.

N. U. Prabhu
I. V. Basawa

Contributors

I. V. BASAWA Department of Statistics, University of Georgia, Athens, Georgia

P. J. BROCKWELL Department of Statistics, Colorado State University, Fort Collins, Colorado

D. J. DALEY Department of Statistics (IAS), Australian National University, Canberra, Australia

MICHAEL DAVIDSEN Copenhagen County Hospital, Herlev, Denmark

R. A. DAVIS Department of Statistics, Colorado State University, Fort Collins, Colorado

J. M. GANI Department of Statistics, University of California at Santa Barbara, Santa Barbara, California

COLIN R. GOODALL School of Engineering and Applied Science, Princeton University, Princeton, New Jersey

P. E. GREENWOOD Department of Mathematics, University of British Columbia, Vancouver, British Columbia, Canada

MARTIN JACOBSEN Institute of Mathematical Statistics, University of Copenhagen, Copenhagen, Denmark

ALAN F. KARR   Department of Mathematical Sciences, The Johns Hopkins University, Baltimore, Maryland

YOUNG-WON KIM   Department of Statistics, University of Georgia, Atlanta, Georgia

IAN W. McKEAGUE   Department of Statistics, Florida State University, Tallahassee, Florida

DAVID M. NICKERSON   Department of Statistics, University of Georgia, Athens, Georgia

MICHAEL J. PHELAN   School of Engineering and Applied Science, Princeton University, Princeton, New Jersey

B. L. S. PRAKASA RAO   Indian Statistical Institute, New Delhi, India

D. A. RATKOWSKY   Program in Statistics, Washington State University, Pullman, Washington

H. SALEHI   Department of Statistics, Michigan State University, East Lansing, Michigan

MICHAEL SØRENSEN   Department of Theoretical Statistics, Institute of Mathematics, Aarhus University, Aarhus, Denmark

TIZIANO TOFONI   Postgraduate School of Telecommunications, L'Aquila, Italy

W. WEFELMEYER   Mathematical Institute, University of Cologne, Cologne, Federal Republic of Germany

LIONEL WEISS   School of Operations Research and Industrial Engineering, Cornell University, Ithaca, New York

Contents

STATISTICAL INFERENCE
IN STOCHASTIC PROCESSES

1
Statistical Models and Methods in Image Analysis: A Survey

Alan F. Karr
The Johns Hopkins University, Baltimore, Maryland

This paper is a selective survey of the role of probabilistic models and statistical techniques in the burgeoning field of image analysis and processing. We introduce several areas of application that engender imaging problems; the principal imaging modalities; and key concepts and terminology. The role of stochastics in imaging is discussed generally, then illustrated by means of three specific examples: a Poisson process model of positron emission tomography, Markov random field image models and a Poisson process model of laser radar. We emphasize mathematical formulations and the role of imaging within the context of inference for stochastic processes.

1 INTRODUCTION

Image analysis and processing is a field concerned with the modeling, computer manipulation and investigation, and display of two-dimensional pictorial data. Like many nascent fields, it is a combination of rigorous mathematics, sound engineering and black magic. Our objec-

Research supported in part by the National Science Foundation under grant MIP-8722463 and by the Army Research Office.

1

tive in this paper is to present some key concepts and issues in image analysis, and to describe some probabilistic/statistical methods applicable to imaging problems. Our emphasis is on mathematical formulations and issues, rather than either theoretical details or empirical techniques.

In the broadest sense an *image* is a function f, whose two-dimensional domain, the *image plane*, is a (compact) subset of $\mathbf{R}^2$. The values of f represent brightness, and may be one-dimensional, in which case the image is called *monochromatic* and its values *gray levels*; or multidimensional, corresponding to brightnesses in several spectral bands, so that the image is colored. For simplicity we deal only with monochromatic images.

The ultimate objective of image analysis is generally—albeit not always—to display an image for viewing, examination and interpretation by humans. Thus, perceptual issues are important; for example, one must be careful not to confuse colored images with "false color" methods for display of monochromatic images. The latter map difficult-to-perceive gray levels onto colors, which are more easily distinguished by human viewers, for the purpose of displaying the image more informatively. We do not deal with perception here, though. The reader should be cautioned, moreover, not to adhere too dogmatically to the visual metaphor, since there are contexts, for example optical character recognition and robotics, in which it is not apparent that this is the best way to think.

In most realistic situations, *images are realizations of random processes*, leading to the following basic definitions.

Definition 1.1. (a) A *continuous image* is the realization of a random field $\{X(x, y) : 0 \le x, y \le 1\}$.

(b) A *digital image* is the realization of a discrete-valued random field $\{X(i, j) : 0 \le i, j \le n\}$. The $X(i, j)$ are referred to as *pixels* (picture elements).

2 SOME GENERALITIES ABOUT IMAGING

Some key forms of imaging are:

CAT: computerized axial tomography
MRI: magnetic resonance imaging
PET: positron emission tomography
OCR: optical character recognition
radar

ladar: laser radar
remote sensing (for example, from satellites)
ultrasound
computer vision and graphics
photography

Common characteristics of imaging problems are as follows:

Digitization. This is necessary in order to analyze and manipulate images by computer, and takes two primary forms. Given an underlying continuous image $\{\tilde{X}(x, y) : 0 \leq x, y \leq 1\}$, an approximating digital image $\{X(i, j) : 0 \leq i, j \leq n\}$ is constructed by

Spatial sampling: The pixels $X(i, j)$ correspond in location to *discrete spatial sampling* of $\tilde{X}$. In most instances the sampling is regular, so that in the absence of quantization $X(i, j) = \tilde{X}(i/n, j/n)$.

Quantization: In a digital system, only finitely many pixel values are possible; these result from quantizing the values of $\tilde{X}$. That is, were there is no spatial sampling, and assuming that $0 \leq \tilde{X} \leq c$ for some constant c, one would have a quantized image $\bar{X}(x, y) = Q(\tilde{X}(x, y))$, where Q is a function from $[0, c]$ to some finite set, say $\{g_0, \ldots, g_k\}$. An example is the function

$$Q(t) = \frac{c}{k}\left[\frac{kt}{c}\right]$$

where $[t]$ is the integer part of t, which yields $k + 1$ equally spaced quantization levels $0, c/k, 2c/k, \ldots, c$.

The final result of both forms of digitization is the digitized image

$$X(i, j) = Q(\tilde{X}(i/n, j/n)) \tag{2.1}$$

Data Transformation. Often in imaging problems the data comprise a mathematical transform of the object of interest. For example, in CAT and MRI, two principal forms of medical imaging, the object is an unknown function $f(x, y)$ representing anatomical structure, and the observed data are (finitely many) line integrals of f:

$$Pf(L) = \int_L f(x, y)\, ds$$

Recovery of f from line integrals is known as the Radon inversion problem. In the case of positron emission tomography (see Section 4), a similar yet different transformation is involved.

Noise. Usually modeled as random, noise may be either inherent to the imaging process or introduced during transmission or storage of data.

Randomness. Randomness enters image analysis in two quite distinct forms, as models of

objects in the image
imaging processes

One objective of this chapter is to argue that the former, in many ways the more important, is rather poorly understood and modeled, whereas the latter is generally well understood and modeled.

Principal issues in image analysis include a number that arise as well in classical, one-dimensional signal processing, such as those that follow. Many of these are "frequency domain" methods in spirit.

Digitization. In order to be processed by computer, a continuous signal must be reduced to digital form, using, for example, the sampling and quantization techniques previously described.

Data Compression. Efficient storage and transmission of image data require that it be represented as parsimoniously as possible. For example, the Karhunen–Loève expansion expresses a stationary random field, without loss of information, in terms of a fixed set of orthonormal functions and a set of uncorrelated random variables. Other techniques for data compression include interpolation and predictive methods, as well as classical concepts from coding theory.

Enhancement. Many signals are subject to known (or estimated) degradation processes, which may be either exogenous, such as non-linear transmission distortion and noise, or endogeneous, such as the "noise" introduced by quantization. Enhancement is intended to remove these effects. For example, "bursty" or "speckle" noise may be removed by smoothing the signal, in other words, subjecting it to a low-pass filter, typically an integral operator. Smoothing removes isolated noise, but at the expense of introducing undesirable blurring. On the other hand, identification of more detailed features may require a high-pass filter (differentiation); for images this can be interpreted as sharpening that removes blur. Smoothing and sharpening techniques are ordinarily implemented in the frequency domain, via multiplication of Fourier transforms. Enhancement may also deal simply with gray levels, for example, in an image degraded by loss of contrast.

Deconvolution.   A linear time-invariant system has the form

$$g(t) = h * f(t) = \int h(t - s)f(s)\, ds \qquad (2.2)$$

where f is the input, g the output, and h the transfer function. Recovery of f from g requires *deconvolution*, effected computationally by means of Fourier transforms.

Filtering.   Depending on the nature of the filter, either large scale or fine scale structure of the signal may be emphasized.

But beyond these are issues truly intrinsic to the two-dimensional and pictorial nature of images:

Restoration.   For images, this constitutes removal of specific kinds of degradations, for example *point degradations* of the form

$$g(x, y) = \int h(x, y; x', y')f(x', y')\, dx'\, dy' \qquad (2.3)$$

where f is the undegraded image, g is the degraded image, and h is the *point spread function*. That is, a point source (a unit intensity source concentrated at a single point, and represented mathematically by the Dirac measure $\epsilon_{(x', y')}$) appears in the degraded image as the measure with density function $h(\cdot, \cdot; x'; y')$. The form of the point spread function depends on the physical process under study. Recovery of f from g in this case typically reduces to a two-dimensional deconvolution problem. Other techniques for restoration include least squares and maximum entropy.

Feature Identification.   Images often contain geometric features that, if not actually defining them, are their most salient attributes. The most important of these are *edges*, which comprise boundaries between different objects in an image, and are characterized by differing pixel values on opposite sides. Edges are high frequency phenomena, and hence techniques to identify them include spatial high-pass filters; more sophisticated means for edge modeling and detection also exist, however, one of which is described in Section 5.

Reconstruction.   In CAT and MRI, as noted above, the measured data are line integrals (see (2.1)) of an unknown function of interest, that one must then reconstruct. A set of such integrals, along lines that are either parallel or emanate in a fan-beam from a single point, is termed a *projection*; the measurements are a set of projections. Most reconstruction procedures are based on the filtered-backprojection algorithm; see (Kak and Rosenfeld, 1982, Chapter 8) for a description.

Matching. Often images of the same object produced at different times or by different imaging modalities must be compared or registered in order to extract as much information from them as possible.

Segmentation. Many images are comprised of large, structured regions, within which pixel values are relatively homogeneous. Such regions can be delineated by locating their boundaries (edge identification), but there are also techniques (some based on statistical classification procedures) that account explicitly for "regional" structure, but deal only implicitly with the boundaries between regions. (The dual viewpoint of region boundaries as both defining characteristics and derived properties of an image is a sometimes frustrating but also powerful aspect of image analysis.)

Texture Analysis. Texture may be regarded as homogeneous local inhomogeneity in images. More broadly, textures are complex visual patterns composed of subpatterns characterized by brightness, colors, shapes, and sizes. Modeling of texture often entails tools such as fractals (Mandelbrot, 1983), while textures play a fundamental role in image segmentation. Mathematical morphology (Serra, 1982) is another key tool.

3 STOCHASTICS IN IMAGING

The role of probability and statistics in imaging is to formulate and analyze models of images and imaging processes.

These include, on the one hand, "classical" degradation models such as

$$Y(x,y) = \int h(x-x'; y-y')X(x',y')\, dx'\, dy' + Z(x,y) \quad (3.1)$$

where X is a random field representing the undegraded image, Y is the degraded image, h is the (spatially homogeneous) point spread function and Z is a noise process, whose effect in the simplest case, as above, is additive. The one-dimensional analog of (3.1) is the time-invariant linear system of (2.2).

On the other hand, a number of "modern" classes of stochastic processes play roles in image analysis:

stationary random fields
Markov random fields
point processes

Image models based on the latter two are treated in the sequel.

In most imaging problems the basic issue is to produce either a literal image (displayed on a computer screen) or a representation suitable for manipulation and analysis. Mathematically, this amounts to *estimation of a two-dimensional function*, which may be either *deterministic and unknown*, so that the statistical problem is one of nonparametric estimation, or *random and unobservable*, in which case the objective is estimation of the realization-dependent value of an unobservable random process (referred to in Karr (1986) as *state estimation*).

Thus, from the stochastic viewpoint, image analysis falls within the purview of inference for stochastic processes, in which there is rapidly increasing activity by probabilists, engineers and statisticians. In the remaining sections of this paper we attempt, using three rather diverse examples, to elucidate the connections between imaging and applications, computation, probability, and statistics.

The literature on statistical models and methods for image analysis is growing steadily; two useful and fairly general sources are Ripley (1988) and Wegman/DePriest (1986).

4 POSITRON EMISSION TOMOGRAPHY

In this section we treat positron emission tomography, one of several complementary methods for medical imaging. We formulate a stochastic model for the process that produces the images and analyze several statistical issues associated with the model. In the process the statistical issues are related to some key inference problems for point processes. The material in this section is based on Karr et al. (1990), to which the reader is referred for proofs and additional details. See also Vardi et al. (1985) for an analogous discrete model; many of the key ideas emanate from this paper.

4.1 The Physical Process

Positron emission tomography (PET) is a form of medical imaging, used to discern anatomical structure and characteristics without resorting to invasive surgical procedures. In PET, metabolism is measured, from which anatomical structure is inferred.

In summary, the physics of PET are as follows (refer to Figure 1.1):

1. Based on the organ to be imaged—for concreteness, the heart (Figure 1.1 represents a transverse planar section), a protein is selected that is known to be metabolized there and not elsewhere.
2. The protein is synthesized in a manner that "tags" it with a radioactive material, often C^{14} or F^{18}; this material, when it decays, emits positrons, which are anti-electrons.
3. The tagged protein is injected into the circulatory system, which transports it to the heart, where it is absorbed because of its role in cardiac metabolism.
4. When a positron is emitted, it is—after traveling a short but nonnegligible distance—annihilated by combining with an electron.
5. The result of a positron/electron annihilation is two photons (γ-rays), that travel in opposite directions (by conservation of momentum) along a randomly oriented straight line.
6. The object being imaged is surrounded by a planar ring of photon detectors; each registers counts of individual photons impinging on it, which are recorded for analysis of the image.

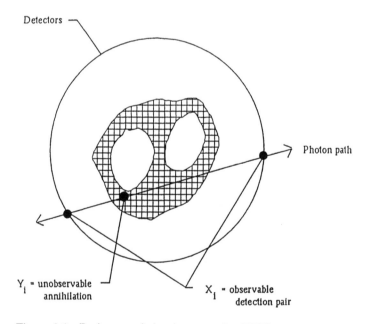

Figure 1.1 Positron emission tomography (PET).

A typical image is based on approximately 10^6 photon-pair detections obtained over a period of several minutes. Often a series of images, offset in the third coordinate from one another, is produced by shifting bodily the ring of detectors; the various two-dimensional images may be analyzed individually or in concert, or combined to form a three-dimensional image.

Returning to a single image, each annihilation occurs within the anatomical structure at a random two-dimensional location Y_i, which *cannot be observed directly*. Instead, one observes a two-dimensional random vector X_i corresponding to the two locations on the ring of detectors at which photons produced by the annihilation are registered. One then knows that Y_i lies along the line segments with endpoints X_i, but nothing more.

Physical issues associated with PET include

Classification: Photon detections at the various detectors must be paired correctly. In practice, misclassification is believed to occur in 30% or more of annihilations.

Noise: Stray photons, not resulting from annihilations, may be registered by the detectors.

Travel distance: Positrons are annihilated near but not at the point of emission.

Reabsorption: As many as 50% of the photons created by annihilations may be absorbed in the body, and hence not detected.

Time-of-flight: Currently available equipment is not sufficiently sophisticated to measure the time differential (it is on the order of 10^{-10} seconds) between the arrivals of the two photons created by a single annihilation. (Were this possible, even within a "noise" level, then Y_i would be known at least approximately.)

Movement: Because the data are collected over a period of several minutes, movement of the subject, either intrinsic, as in the case of the heart, or spurious, may occur.

Despite the errors introduced by these phenomena, they are ignored in the simple model we present below.

4.2 The Stochastic Model

There is abundant physical and mathematical basis for assuming, as we do, that the two-dimensional point process

$$N^* = \sum \epsilon_{Y_i} \qquad (4.1)$$

of positron/electron annihilations is a Poisson process (on the unit disk

$D = \{x \in \mathbf{R}^2 : |x| \leq 1\}$) whose *unknown* intensity function α represents indirectly, via metabolic activity, the anatomical structure of interest, in our example the heart.

We further assume, also with sound basis, that

The X_i are conditionally independent given N^*
The conditional distribution of X_i, given N^*, depends only on Y_i, via a *point spread function k*:

$$P\{X_i \in dx \mid N^*\} = k(x \mid Y_i) \, dx \qquad (4.2)$$

That is, $k(\cdot \mid y)$ is the conditional density of the observed detection location corresponding to an annihilation at y; the randomness it introduces represents the random orientation of the photons engendered by the annihilation, and to some extent travel distances.

It follows that the marked point process

$$\bar{N} = \sum \epsilon_{(Y_i, X_i)} \qquad (4.3)$$

which is obtained from N^* by independent marking (Karr, 1986, Section 1.5), is a Poisson process with intensity function

$$\eta(y, x) = \alpha(y) k(x \mid y) \qquad (4.4)$$

In consequence, the point process

$$N = \sum \epsilon_{X_i} \qquad (4.5)$$

of observed detections is also Poisson, with intensity function

$$\lambda(x) = \int \eta(y, x) \, dy = \int k(x \mid y) \alpha(y) \, dy \qquad (4.6)$$

The goal is to estimate the intensity function α of the positron/electron annihilation process. However, before discussing this in detail, we consider some generalities associated with inference for Poisson processes.

4.3   Estimation of Intensity Functions of Poisson Processes

Consider the problem of estimating the intensity function of a Poisson process N on $\mathbf{R}^d$ from observation of i.i.d. copies $N_1, \ldots, N_n$. Denoting by P_λ the probability law of the observations N_i when the unknown intensity function is λ, the relevant log-likelihood function is (Karr, 1986, Proposition 6.11) given by

$$L(\lambda) = \log\left(\frac{dP_\lambda}{dP_0}\right) = \int (\lambda_0 - \lambda) \, dx + \int \log(\lambda/\lambda_0) \, dN'' \qquad (4.7)$$

where P_0 is the probability law corresponding to a fixed function λ_0 and $N^n = \Sigma_{i=1}^n N_i$, which is a sufficient statistic for λ given the data $N_1, \ldots, N_n$.

This log-likelihood function of (4.7) is unbounded above (except in the uninteresting case that $N^n(\mathbf{R}^d) = 0$) and maximum likelihood estimators do not exist. We next introduce a technique that circumvents this difficulty in an especially elegant manner.

4.4 The Method of Sieves

The method of sieves of U. Grenander and collaborators is one means for "rescuing" the principle of maximum likelihood in situations where it cannot be applied directly, typically because the underlying parameter space is infinite-dimensional. We begin with a brief general description.

Suppose that we are given a statistical model $\{P_\xi; \xi \in I\}$, indexed by an infinite-dimensional metric space I, and i.i.d. data $X_1, X_2, \ldots$. Then, in the absence of special structure we assume the n-sample log-likelihood functions $L_n(\xi)$ are unbounded above in ξ.

Grenander's idea is to restrict the index space but let the restriction become weaker as sample size increases, all in such a manner that maximum likelihood estimators not only exist within the sample-size-dependent, restricted index spaces but also fulfill the desirable property of consistency. More precisely, the paradigm (Grenander, 1981) is as follows:

1. For each $h > 0$ define a "compact" subset $I(h)$ of I such that for each n and h there exists an n-sample maximum likelihood estimator $\hat{\xi}(n, h)$ *relative to* $I(h)$:

$$L_n(\hat{\xi}(n, h)) \geq L_n(\xi) \qquad \xi \in I(h) \qquad (4.8)$$

2. Construct the $I(h)$ in such a manner that as h decreases to 0, $I(h)$ increases to a dense subset of I.
3. Choose *sieve mesh* h_n to depend on the sample size n in such a manner that $\hat{\xi}(n, h_n) \to \xi$ under P_ξ, in the sense of the metric on I.

In most applications I is a function space and h has the physical interpretation of a constraint on the "roughness" of functions in $I(h)$. As h decreases, rougher and rougher functions are allowed in $I(h)$. See Karr (1986, 1987), as well as Section 4.5 below, for additional discussion.

Figure 1.2 provides a pictorial interpretation of the method of sieves. There, the squares denote the elements of $I(h)$ and the squares and circles the elements of $I(h')$, where $h' < h$; ξ is the true value of the unknown parameter.

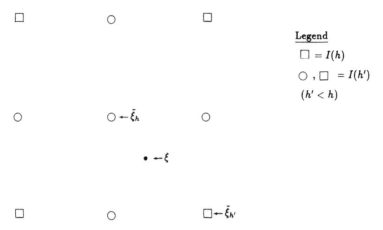

Figure 1.2 The method of sieves.

The reasoning that underlies the method of sieves is as follows:

1. For each h, there exists within $I(h)$ a "regularization" $\tilde{\xi}_h$ of ξ that, if not a best approximation to ξ from $I(h)$, is at least a sufficiently good approximation.
2. Maximum likelihood estimators for a fixed sieve mesh h should satisfy standard asymptotics, so that under the probability P_{ξ_h}, $\hat{\xi}(n, h) \to \tilde{\xi}_h$ as $n \to \infty$ (with, we emphasize, h held fixed).
3. Assuming that $h = h_n$ is chosen to depend on n in such a manner that $h_n \to 0$ as $n \to \infty$, then $\tilde{\xi}_{h_n} \to \xi$ by construction of the sieve.
4. "Hence" provided that $h_n \to 0$ slowly enough, it should be true that $\hat{\xi}(n, h_n) \to \xi$ under P_ξ.

The fundamental tradeoff involved in determining the h_n is esssentially *bias against variance*: as h becomes smaller,

The regularization $\tilde{\xi}_h$ is a better approximation to ξ and hence less biased,

but

ξ_h is also rougher and hence more difficult to estimate, because estimators have higher variance.

In order that estimators $\hat{\xi}(n, h_n)$ be consistent, the h_n must be chosen to balance the competing goals of eliminating both variance (in order that estimators converge) and bias (so that they converge to the right thing).

We now apply the method of sieves to the problem of estimating the intensity function of a Poisson process.

4.5 Application to Inference for Poisson Processes

Our setting is again the following: $N_1, N_2, \ldots$ are i.i.d. Poisson processes on $\mathbf{R}^d$ with unknown intensity function λ satisfying

$\int_{\mathbf{R}^d} \lambda(x) \, dx = 1$.
supp(λ) is compact (albeit perhaps unknown).

Provided that $\int \lambda < \infty$, the former assumption entails no loss of generality, since given the data $N_1, \ldots, N_n$, $\int \lambda$ admits the maximum likelihood estimator $N^n(\mathbf{R}^d)/n$; these estimators are strongly consistent and even asymptotically normal. The latter does involve some restriction mathematically, but none in terms of applications.

The goal is to estimate λ consistently.

The log-likelihood function (omitting terms not depending on λ) is

$$L_n(\lambda) = \int (\log \lambda) \, dN^n \qquad (4.9)$$

which remains unbounded above in λ.

We invoke the method of sieves, using the *Gaussian kernel sieve*. For each $h > 0$, define

$$k_h(x \mid y) = \frac{1}{(2\pi h)^{d/2}} \exp\left[\frac{-\|x - y\|^2}{2h}\right]$$

that is, $k_h(x \mid y) = f(x - y)$, where f is a circularly symmetric Gaussian density with mean zero and covariance matrix hI. The sieve $I(h)$ consists of all intensity functions λ having the form

$$\lambda(x) = \int k_h(x \mid y) F(dy) \qquad (4.10)$$

where F is a probability measure on $\mathbf{R}^d$. Elements of $I(h)$ are, in particular, smooth densities on $\mathbf{R}^d$. See Geman and Hwang (1982) for an application of the same sieve to estimation of probability density functions.

Restricted to $I(h)$, viewed as indexed by the family $\mathscr{P}$ of probabilities on $\mathbf{R}^d$, the log-likelihood function L_n of (4.9) becomes

$$L_n(F) = \int \left[\log \int k_h(x \mid y) F(dy)\right] N^n(dx) \qquad (4.11)$$

The inference issues now become

existence of maximum likelihood estimators in $I(h)$
computation of maximum likelihood estimators in $I(h)$
consistency of the sieve maximum likelihood estimators

We shall discuss these in turn.

Existence of maximum likelihood estimators in the $I(h)$ requires only that $k(x \mid y) \to 0$ as $\|x\|$ or $\|y\| \to \infty$.

Proposition 4.1. For each n and h, there exists $\hat{F}(n, h) \in I(h)$ such that

$$L_n(\hat{F}(n, h)) \geq L_n(F) \tag{4.12}$$

for all $F \in I(h)$. Moreover, $\hat{F}$ is supported by the convex hull of the points of N^n.

From (4.12) one defines an estimator of the intensity function by substitution:

$$\hat{\lambda}(n, h)(x) = \int k_h(x \mid y)\hat{F}(n, h)(dy) \tag{4.13}$$

Computation of maximum likelihood estimators is a more delicate issue. We have sought to use the EM algorithm (Dempster et al., 1977; Wu, 1983), but the space $\mathcal{P}$ is infinite-dimensional and some convergence issues remain unresolved. Nevertheless, we relate what is known regarding our problem.

The EM algorithm is an iterative method for calculating maximum likelihood estimators whose direct computation is difficult. We present a brief general description.

Consider a random vector (Y, X) governed by a statistical model $\{P_\theta : \theta \in \Theta\}$, in which X is observable, but Y is not. Define the log-likelihood functions

$$L(\theta, x) = \log P_\theta\{X = x\}$$
$$\Lambda(\theta, y) = \log P_\theta\{Y = y\}$$

The goal is to calculate, from the observations X, a maximum likelihood estimator $\hat{\theta}$ of θ. Of course, since only $L(\cdot, X)$ can be calculated from the observations, $\hat{\theta}$ must maximize it.

In a variety of situations, for example, mixture models (Lindsay, 1983a, 1983b), maximization of $\Lambda(\cdot, Y)$ is easy, either numerically or perhaps even in closed form, but maximization of $L(\cdot, X)$ is much more difficult. The EM algorithm seeks to ameliorate this situation by iteration. Given a "previous" estimated value $\hat{\theta}_{\text{old}}$, the "new" estimated value is given by

$$\hat{\theta}_{\text{new}} = \arg\max_{\theta} E_{\hat{\theta}_{\text{old}}}[\Lambda(\theta, Y) \mid X] \qquad (4.14)$$

There is no guarantee, of course, that this problem is more amenable to a solution than that for L, but the hope is that the function $E_{\hat{\theta}_{\text{old}}}[\Lambda(\cdot, Y) \mid X]$, inherits the computational tractability of Λ. In many specific cases, this turns out to be so. Note that since it is a conditional expectation given X, this function is computable from the observations.

Thus, given an initial estimate $\hat{\theta}_{\text{init}}$, the EM algorithm generates a sequence $(\hat{\theta}_{(k)})$ of estimated values. The iteration (4.14) increases the value of L:

$$L(\hat{\theta}_{\text{new}}) \geq L(\hat{\theta}_{\text{old}})$$

and the inequality is strict unless $\hat{\theta}_{\text{new}} = \hat{\theta}_{\text{old}}$. Moreover, provided that certain regularity conditions be fulfilled, in particular that Θ be finite-dimensional, then $\hat{\theta}_{(k)} \to \hat{\theta}$, the maximizer of $L(\cdot, X)$. See Dempster et al. (1977) and Wu (1983) for details.

Returning to computation of sieve maximum likelihood estimators for Poisson intensity functions, we have the following result.

Proposition 4.2. (a) The iteration

$$\hat{F}_{\text{new}}(dy) = \hat{F}_{\text{old}}(dy) \int \frac{k_h(x \mid y)}{\int k_h(x \mid z)\hat{F}_{\text{old}}(dz)} \frac{N^n(dx)}{N^n(\mathbf{R}^n)} \qquad (4.15)$$

corresponds to an EM algorithm for maximizing $L_n(F)$.

(b) Provided that the support of $\hat{F}_{\text{init}}$ is $\mathbf{R}^d$, the sequence $(\hat{F}_{(k)})$ of iterates produced by (4.15) is tight.

As yet, however, we have not shown that the EM sequence $(\hat{F}_{(k)})$ converges, nor (even assuming that it does) that the limit maximizes L_n.

An alternative approach to computation of the maximum likelihood estimators $\hat{F}(n, h)$ relies on showing that these estimators have finite support. Suppose that $N^n = \sum_{i=1}^{M} \epsilon_{X_i}$.

Theorem 4.3. For each n and h, the sieve maximum likelihood estimators $\hat{F} = \hat{F}(n, h)$ satisfy

$$\frac{1}{M} \sum_{i=1}^{M} \frac{k_h(X_i - v)}{\int k_h(X_i - w)\hat{F}(dw)} = 1 \qquad v \in \text{supp } \hat{F} \qquad (4.16)$$

$$\frac{1}{M} \sum_{i=1}^{M} \frac{k_h(X_i - v)}{\int k_h(X_i - w)\hat{F}(dw)} \le 1 \qquad \text{for all } v \qquad (4.17)$$

Consequently, the support of $\hat{F}$ is finite.

The conditions (4.16)–(4.17) correspond, respectively, to the first and second Kuhn–Tucker conditions for the problem of maximizing the log-likelihood function (4.11).

The characterization in Theorem 4.3 engenders a finite-dimensional EM algorithm for computation of $\hat{F}$. Denote by $\mathcal{P}_q$ the set of probability distributions on $\mathbf{R}^d$ whose support consists of at most q points (so that every $F \in \mathcal{P}_q$ admits a representation $F = \sum_{j=1}^{q} p(j)\epsilon_{v(j)}$).

Theorem 4.4. For $q \ge M$, within $\mathcal{P}_q$ the iteration that converts $\hat{F}_{\text{old}} = \sum_{j=1}^{q} p(j)\epsilon_{v(j)}$ into $\hat{F}_{\text{new}} = \sum_{j=1}^{q} p'(j)\epsilon_{v'(j)}$, where

$$p'(j) = p(j)\frac{1}{M} \sum_{i=1}^{M} \frac{k_h(X_i - v(j))}{\sum_{l=1}^{q} p(l)k_h(X_i - v(l))} \qquad (4.18)$$

$$v'(j) = \frac{\sum_{i=1}^{M} X_i \{k_h(X_i - v(j))/\sum_{l=1}^{q} p(l)k_h(X_i - v(l))\}}{\sum_{i=1}^{M} \{k_h(X_i - v(j))/\sum_{l=1}^{q} p(l)k_h(X_i - v(l))\}} \qquad (4.19)$$

is an EM algorithm.

Known convergence theory for finite-dimensional EM algorithms (Wu, 1983) applies to the algorithm of Theorem 4.4.

Finally, consistency of maximum likelihood estimators $\hat{\lambda}$ of (4.13), with respect to the L^1-norm, is established in the following theorem.

Theorem 4.5. Suppose that
  (a) λ is continuous with compact support.
  (b) $|\int \lambda(x) \log \lambda(x)\, dx| < \infty$.
Then for $h_n = n^{-1/8 + \epsilon}$, where $\epsilon > 0$,

$$\|\hat{\lambda}(n, h_n) - \lambda\|_1 \to 0 \qquad (4.20)$$

almost surely.

4.6   Application to PET

The connection between PET and our discussion of inference for Poisson processes is that estimation within a sieve set $I(h)$ is the same problem mathematically as computation of the maximizer of the likelihood function for the PET problem, given (compare (4.7)) by

$$L(\alpha) = \int \left(\lambda_0(x) - \int k(x \mid y)\alpha(y)\, dy \right) dx$$

$$+ \int \log\left(\frac{\int k(x \mid y)\alpha(y)\, dy}{\lambda_0(x)} \right) N(dx)$$

More succinctly, for PET, the intensity function has the form

$$\lambda_\alpha(x) = \int k(x \mid y)\alpha(y)\, dy$$

whereas in connection with the method of sieves,

$$\lambda_F(x) = \int k_h(x \mid y)F(dy)$$

In both cases the point spread function k is known.

Proposition 4.2 and an analog of Theorem 4.3 apply to the PET problem.

5 MARKOV RANDOM FIELD IMAGE MODELS

The image models described in this section differ substantially from the Poisson process model of the preceding section: they are based on random fields (stochastic processes with multidimensional parameter sets); they are inherently discrete; they incorporate dependence among pixel values by means of auxiliary stochatic processes; and they admit almost unlimited modeling flexibility. Of course, all this does not come unfettered, and applicability in any specific situation must still be determined by resort to physical principles.

The principal source for the material in this section is Geman and Geman (1984), to which the reader is referred for details. See also Besag (1974, 1986); Cross and Jain (1983) contains an application of Markov random fields to the modeling and analysis of texture.

5.1 The Model

The image is modeled as a random field $X = \{Y_p, p \in S; Z_l, l \in L\}$, where

Y is an *observable pixel process*, indexed by a set S of sites.

Z is an *unobservable line process*, indexed by a set L of lines.

The random variables Y_p and Z_l take their values in a finite state space E.

As illustrated in examples below, the role of lines is to introduce additional dependence into the pixel process in a controlled manner. Thus, on the one hand the line process is a modeling tool of enormous power. On the other hand, however, it is artificial in that the "lines" need not (and generally do not) represent objects that exist physically, but are instead artifacts of the model. In a sense, this dilemma is merely one manifestation of the somewhat schizophrenic view of "edges" that seems to prevail throughout image analysis: are edges defined by the image, or do they define it?

Returning to the model, the crucial mathematical assumption is that X is a Markov random field. That is, let $G = S \cup L$, and assume that there is defined on G a symmetric *neighborhood structure*: for each α, there is a set $N_\alpha \subseteq G$ of *neighbors* of α, with $\beta \in N_\alpha$ if and only if $\alpha \in N_\beta$. Then, X is a *Markov random field* if

(a) $P\{X = x\} > 0$ for each configuration $x \in \Omega = E^G$,
(b) For each α and s,

$$P\{X_\alpha = s \mid X_\beta, \beta \neq \alpha\} = P\{X_\alpha = s \mid X_\beta, \beta \in N_a\} \qquad (5.1)$$

That is, once the values of X at the neighbors of α are known, additional knowledge of X at other sites is of no further use in predicting the value at α.

An equivalent statement is that the probability $P\{X \in (\cdot)\}$ is a *Gibbs distribution*, in the sense that

$$P\{X = x\} = \frac{1}{Z_T} \exp\left[-\frac{1}{T} \sum_C V_C(x) \right] \qquad (5.2)$$

where T is a positive parameter with the physical interpretation of temperature, the summation is over all cliques C of the graph G, and for each C, the function $V_C(x)$ depends on x only through $\{x_\alpha : \alpha \in C\}$. (A subset C of G is a *clique* if each pair of distinct elements of C are neighbors of one another. Note that this definition is satisfied vacuously by singleton subsets of C.) The V_C and the function

$$U(x) = \sum_C V_C(x) \qquad (5.3)$$

particularly the latter, are termed *energy functions*.

Interpretations are as follows:

The energy function determines probabilities that the system assumes various configurations: the lower the energy of the configuration x, the higher the probability $P\{X = x\}$.

The temperature modulates the "peakedness" of the probability distribution $P\{X = (\cdot)\}$: the lower T, the more this distribution concentrates its mass on low energy configurations.

The function of T defined by

$$Z_T = \sum_x e^{-U(x)/T} \tag{5.4}$$

which plays the role of a normalizing constant in (5.2), is known on the basis of analogies to statistical mechanics as the *partition function*. Its computation is notoriously difficult.

That (5.1) and (5.2) are equivalent is an important and nontrivial property; see Kindermann and Snell (1980) for a proof.

One observes only the *degraded image*, a random field $\tilde{Y} = \{\tilde{Y}_p : p \in S\}$ indexed solely by pixels and having the form

$$\tilde{Y}_p = \phi(H(Y)_p) + \nu_p \tag{5.5}$$

where

H is a known blurring function, which operates on the entire undegraded pixel process Y.

ϕ is a known, possibly nonlinear, transformation, applied pixel-by-pixel.

ν is (typically, white Gaussian) noise. Though the noise in (5.5) is shown as additive, it need not be. In essentially all cases it is independent of X.

Note that the line process Z does not appear in (5.5).

Although in principle the value of the blurred image $H(Y)$ at a given pixel can depend on the values of Y at every pixel, typically H is "local" in nature. For example, one may model blur as local averaging, a phenomenon well-known in image analysis (it may even be introduced deliberately, in order to remove certain forms of noise), in which each pixel value is replaced by a weighted average of its value and those of neighboring pixels:

$$H(Y)_p = \gamma Y_p + (1 - \gamma) \frac{1}{|N_p|} \sum_{q \in N_p} Y_q$$

where $\gamma \in (0, 1)$ and N_p is the set of pixel neighbors of p.

The goal of the analysis is to recover the original image X, although usually only the pixel process Y is of interest. Note that the unknown function X is random in this case, rather than, as in Sections 4 and 6, a statistical parameter. Thus, our context is *state estimation*, the realization-dependent reconstruction of an unobservable random process. This is effected via Bayesian techniques: the reconstructed image $\hat{X}$ is the *maximum a priori probability*, or MAP, estimator: that value which maximizes the conditional probability $P\{X = (\cdot) \mid \tilde{Y}\}$, where $\tilde{Y}$ is the degraded image of (5.5). Thus, the reconstructed image is given by

$$\hat{X} = \arg \max_x P\{X = x \mid \tilde{Y}\} \qquad (5.6)$$

The central issues are

Model specification, that is, choice of
  the pixel index (site) set S
  the line set L
  the neighborhood structure on $G = S \cup L$
  the (prior) distribution $P\{X \in (\cdot)\}$ or equivalently the energy function
  U of (5.3)
Computation of conditional probabilities
Solution of a difficult, unstructured, discrete optimization problem in
  order to calculate the MAP estimator $\hat{X}$

The scale of the problem can be massive: in realistic situations, the image is $256 \times 256 = 65536$ pixels, there are 16 grey levels, and hence the pixel process alone admits 16^{65536} configurations. In particular, solution of the MAP estimation problem by nonclever means is out of the question.

5.2 An Example

We now illustrate with perhaps the simplest example. Suppose that

$S = \mathbf{Z}^2$, the two-dimensional integer lattice.
$L = D^2$, the dual graph of S, as depicted in Figure 1.3.
Neighbors in S and L are defined as follows:
  Within S: pixels $p = (i, j)$ and $r = (h, k)$ are neighbors if they are
  physical neighbors, that is, if $|i - h| + |j - k| = 1$.
  Within L: lines l and m are neighbors if they share a common
  "endpoint" (which is *not* a site of the pixel process); thus each of
  the four lines in Figure 1.4 is a neighbor of the others.

Legend

$\bigcirc$ = elements of S

$|$ = elements of L

Figure 1.3 The sets S and L.

Figure 1.4   Neighboring lines.

Between S and L: a pixel p and line l are neigbhbors if they are physically adjacent.
The Y_p and Z_l take values in $\{-1, 1\}$.

The cliques and (typical) energy functions for this example are as follows:

Pixels: The only cliques composed solely of pixels are singletons and adjacent pairs. The pixel-pair energy function given by

$$V_C(\{y_p, y_r\}) = b y_p y_r$$

where b is a constant, corresponds to the classical Ising model. If $b > 0$ configurations in which adjacent pixels differ have lower energy, while if $b < 0$, configurations having like adjacent pixels are more likely.

Lines: Within rotations and reflections, cliques composed only of lines are of the five varieties in Figure 1.5. The energy function there illustrates the role of the line process in image modeling; it favors (via lower energy, hence higher probability) long lines, as opposed to corners and short or intersecting lines.

Mixed: Up to rotation, the cliques containing both pixels and lines are all of the form shown in Figure 1.6. The energy function

$$V_C(\{y_p, y_r, z_l\}) = 0 \qquad z_l = -1$$
$$= 1 \qquad z_l = 1, y_p = y_r$$
$$= -1 \qquad z_l = 1, y_p \neq y_r$$

for the clique $C = \{y_p, y_r, z_l\}$, demonstrates as well the role of the

Clique	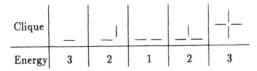				
Energy	3	2	1	2	3

Figure 1.5   Energies of line cliques.

Figure 1.6 The "only" mixed clique.

line process. Lines are either present or not, and if present represent "edges," that is, boundaries between different regions of the image. Thus, if $z_l = -1$, there is no edge at l and the energy of C is independent of the values of y_p and y_r. On the other hand, if $z_l = 1$, there is an edge at l and the energy of C is 1 or -1 accordingly as $y_p = y_r$ or $y_p \neq y_r$, so that the lower energy — and hence higher probability — corresponds to the case that pixel values differ across l.

5.3 Characterization of the Degraded Image

We return to the model presented in Section 5.1.

Theorem 5.1. If the blurring function H is "local" and the noise is white Gaussian with parameters μ, σ^2, then for each $\tilde{y}$, $P\{X = (\cdot) \mid \tilde{Y} = \tilde{y}\}$ is a Gibbs distribution with respect to a particular neighborhood structure on G and with energy function

$$\tilde{U}(x) = U(x) + \|\mu - (\tilde{y} - \phi(H(y)))\|^2/2\sigma^2 \qquad (5.7)$$

Hence the MAP estimator is the minimizer of the energy function $\tilde{U}$, so that solution of the MAP estimation problem entails minimization of the energy function of a Markov random field.

5.4 Energy Function Minimization

Let X be a Markov random field on a finite graph (with vertex set) G, whose energy function U is given by (5.3). The goal is to find a configuration x^* minimizing U:

$$U(x^*) \leq U(x) \qquad (5.8)$$

for all x. That this problem admits a (not necessarily unique) solution is not at issue; the space Ω, albeit possibly of immense cardinality, is finite. The difficulty is numerical computation of x^*.

Given the energy function U, for each temperature $T > 0$ define the probability distribution

$$\pi_T(x) = \frac{1}{Z_T} e^{-U(x)/T} \tag{5.9}$$

where $Z_T = \Sigma_x\, e^{-U(x)/T}$ is the partition function.

Minimization of U is based on two key observations, the first stated as a proposition.

Proposition 5.2. As $T \to 0$, π_T converges to the uniform distribution π_0 on the set of global minima of U.

Moreover, for fixed T, π_T can be computed by "simulation," which requires only particular conditional probabilties and does not entail computation of the partition function Z_T.

To make this more precise, we introduce the local characteristics of a Gibbs distribution π with enegy function U. Let X be a Markov random field satisfying $P\{X = x\} = \pi(x)$, and for $\alpha \in G$, $x \in \Omega$ and $s \in E$ (the state space of X) define $\tau(\alpha, x; s) \in \Omega$ by

$$\tau(\alpha, x; s)_\beta = s \qquad \beta = \alpha$$
$$= x_\beta \qquad \beta \neq \alpha$$

That is, $\tau(\alpha, x; s)$ coincides with x except (possibly) at α, where it takes the value s.

Definition. The *local characteristics* of π are the conditional probabilities

$$\begin{aligned}
L_\pi(\alpha, x; s) &= P\{X_\alpha = s \mid X_\beta = x_\beta, \beta \neq \alpha\} \\
&= \frac{\pi(\tau(\alpha, x; s))}{\Sigma_v\, \pi(\tau(\alpha, x; v))} = \frac{e^{-U(\tau(\alpha,x;s))/T}}{\Sigma_v\, e^{-U(\tau(\alpha,x;v))/T}}
\end{aligned} \tag{5.10}$$

Note that the partition function cancels from the ratios that define the local characteristics.

The next step is to construct an inhomogeneous Markov chain whose limit distribution is π. Let (γ_k) be a sequence of elements of G, and let $(A(k))$ be a stochastic process with state space Ω and satisfying

$$P\{A(k + 1) = \tau(\gamma_{k+1}, x; s) \mid A(k) = x\} = L_\pi(\gamma_{k+1}, x; s) \tag{5.11}$$

This process evolves as follows: from time k to $k + 1$, only the value of A at γ_{k+1} changes, and does so according to the local characteristics of π. That A is a Markov process is evident from (5.11); less apparent is the following result, which states that its limit distribution is π.

Theorem 5.3. If for each $\beta \in G$, $\gamma_k = \beta$ for infinitely many values of k, then regardless of $A(0)$,

$$\lim_{k \to \infty} P\{A(k) = x\} = \pi(x), \qquad x \in \Omega \qquad (5.12)$$

Simulation of the process A, as described in detail in Geman and Geman (1984), is simple and straightforward. In particular,

Only the local characteristics of π are involved, and hence computation of the partition function is not required.

The local characteristics simplify, since the conditional probability in (5.10) is equal to $P\{X_\alpha = s \mid X_\beta = x_\beta, \beta \in N_\alpha\}$.

Only one component of A is updated at each time.

Implementation of the updating scheme can be realized on a parallel computer in which each processor communicates only with its neighbors in the graph G.

Returning to the problem of minimizing the energy function U, for each temperature T define π_T via (5.9). The method of *simulated annealing* effects minimization of U by carrying out the processes of Proposition 5.2 and Theorem 5.3 simultaneously, as we explain shortly. The analogy is to the physical process of annealing, whereby a warm object cooled sufficiently slowly retains properties that would be lost if the cooling occurred too rapidly. An alternative interpretation is that simulated annealing permits the process A constructed below to escape from the local minima of U that are not global minima; for details of this interpretation and additional theoretical analysis, see Hajek (1988).

The key idea is to let the temperature depend on time, and to converge to zero as $k \to \infty$. More precisely, let (γ_k) be an updating sequence as above, let $T(k)$ be positive numbers decreasing to zero, and now let A be a process satisfying

$$P\{A(k + 1) = \tau(\gamma_{k+1}, x; s) \mid A(k) = x\} = L_{\pi_{T(k)}}(\gamma_{k+1}, x; s) \quad (5.13)$$

In words, the transition of A at time k is driven by the Gibbs distribution $\pi_{T(k)}$.

The crucial theoretical property is that — provided the updating scheme is sufficiently regular and the temperature does not converge to zero too quickly — the limit distribution of A is π_0.

Theorem 5.4. Assume that there exists $l \geq |G|$ such that for all k, $G \subseteq \{\gamma_{t+1}, \ldots, \gamma_{t+l}\}$, and denote by Δ the difference between the maximum and minimum values of U. Then provided that $(T(k) \to 0$ and)

$$T(k) \geq \frac{|G|\Delta}{\log k} \tag{5.14}$$

we have

$$\lim_{k \to \infty} P\{A(k) = x\} = \pi_0(x) \tag{5.15}$$

for each x.

In practice, not surprisingly, the rate entailed by (5.14), though sharp, is much too slow; successful empirical results have been reported for much more rapid cooling schedules.

6 THE GEOMETRY OF POISSON INTENSITY FUNCTIONS

In this section we discuss a model of laser radar, or *ladar*, proposed in Karr (1988). As the name suggests, ladar is a form of ranging and detection based on reflection, but the incident energy is provided by a laser rather than a radio transmitter. This variant of classical radar is less subject to certain forms of interference, and has been suggested for use in a variety of contexts, one of which is identification of "targets" in a scene with a complicated background (for example, a battlefield). Our model constitutes a very simple approach to this complicated problem.

6.1 The Model

Consider an image composed of relatively bright objects arrayed against a dark background. The objects have known geometric characteristics — size and shape — belonging to a finite catalog of object types. The image is to be analyzed with the objective of determining

the kinds and numbers of objects present
the locations of the objects

In this section we present a simplified probabilistic model of the image and imaging process, and some empirical investigations of its statistical properties.

Specifically, we suppose that the imaging process (as is ladar) is electromagnetic or optical, so that differing brightnesses correspond to differing "spatial" densities of photon counts. Let $E = [0, 1] \times [0, 1]$, representing the image itself rather than the scene being imaged. In target or character recognition problems, the image consists of a set S of objects of interest (targets)

a background $E \backslash S$

The objects are brighter than the background.

The set of targets has the form

$$S = \bigcup_{i=1}^{k} \left((A_i + t_i) \cap E \right) \qquad (6.1)$$

where k is a nonnegative integer, the A_i are elements of a finite set $\mathfrak{A}$ of subsets of E ($\mathfrak{A}$ is the "object catalog"), and the t_i are elements of E. (For $A \subseteq E$ and $t \in E$, $A + t = \{y + t : y \in A\}$.) (In principle, k, the A_i and the t_i are random, but here we take them as deterministic.) There is no assumption that the objects $A_i + t_i$ are disjoint, so that partial or even total occlusion of one object by others is possible. Also, varying orientations of the objects are not permitted except insofar as they are incorporated in the object catalog $\mathfrak{A}$.

Based on Poisson approximation theorems for point processes it is reasonable to suppose the image is a Poisson process whose intensity function λ assumes one value on the target set S and a smaller value on the background $E \backslash S$:

$$\lambda(x) = b + a \qquad x \in S$$
$$= b \qquad x \in E \backslash S$$

where b, the "base rate," and a, the "added rate," are positive constants, possibly unknown.

Analysis of the image requires that one address the following statistical issues:

estimation of k, the A_i and the t_i, the statistical parameters of the model

tests of hypotheses regarding k, the A_i and the t_i

estimation of the intensity values a and b

In this section, we consider the following sample cases:

1. S consists of a single square with sides of length 0.1 and parallel to the x- and y-axes, whose center is at a fixed but unknown location (within the set $[.05, .95] \times [.05, .95]$).
2. S consists of two (not necessarily disjoint) squares with sides of length 0.1 parallel to the axes, and with centers at fixed but unknown locations.
3. S consists with probability 1/2 of one square with sides of length 0.1 and parallel to the axes, and with center at a fixed but unknown

location, and with probability 1/2 of one disk of radius 0.0564, with center at a fixed but unknown location (in [.0564, .9436] × [.0564, .9436]).

(The restrictions on the centers are in order that S lie entirely within E.) We derive maximum likelihood estimators and likelihood ratio tests and use simulation results to assess their behavior.

In all cases a and b are taken to be known. Although our results to date are incomplete, we have tried also to address the effects of the absolute and relative values of a and b on our statistical procedures.

6.2 Squares with Unknown Locations

We consider first the case where the target consists of a single square S with sides of length 0.1 parallel to the axes, and whose center has fixed but unknown location in [.05, .95] × [.05, .95]. The observed data are the points, random in number and location, of the point process N.

The likelihood function associated with this problem is given by

$$L(S) = -(a + b)|S| - b|E \backslash S|$$
$$+ N(S)\log(a + b) + N(E \backslash S)\log b \qquad (6.3)$$

where $|A|$ denotes the area of the set A. Since $|S|$ is fixed at 0.01, it is evident that L is maximized at the value $\hat{S}$ that maximizes $N(S)$, that is, for that square containing the largest number of points of the process N. In particular, $\hat{S}$ is not unique, since $N(S)$ does not vary under minor perturbations of S; however, computation of a maximizer $\hat{S}$ can performed via the algorithm implicit in the following theorem.

Theorem 6.1. With probability one, there is a maximizer $\hat{S}$ of L that contains two points of N on adjacent sides of its boundary.

The complexity of the algorithm is $O(N(E)^3)$, since $O(N(E)^2)$ candidate squares S must be checked and $N(S)$ must be determined for each.

We have performed rather extensive numerical simulations of the process and computational algorithm, corresponding primarily to different values of the relative brightness $r = (a + b)/b$, with the base level b held constant at 200. The following table summarizes our results. In it, the target square is considered to have been located correctly if both coordinates of the center of the maximum likelihood estimator are within 0.05 of those of its center.

a	r	Cases	No. correct	% correct
1000	6.0	22	20	90.9
500	3.5	50	19	38.0
250	2.25	50	9	18.0
125	1.625	55	5	9.1
50	1.25	50	1	2.0

The centers of the "true" squares varied case-by-case, and were generated by Monte Carlo methods.

The following are the key features of the data.

1. The accuracy with which the square target is located varies dramatically as a function of the additional brightness a:

When $a = 1000$, the accuracy is very high; in only two of the 22 cases is there an error.

Even when $a = 500$, accuracy degenerates substantially, with gross errors in 31 of 50 cases.

For $a = 250$, let alone the lower values of 125 and 50, there is effectively no capability to locate the target. In only 9 of the 50 cases when $a = 250$ is the estimated location near to the actual; the corresponding numbers are 5/55 for $a = 125$ and 1/50 for $a = 50$.

2. From realization to realization, the accuracy varies depending on the (random) number of added points in the square. When $a = 1000$ the mean number of added points is 10, whereas when $a = 50$ it is 0.5. For the case that $a = 1000$, both of the gross errors correspond to relatively low numbers of added points. Similar associations may be observed in the other tables as well. By the time that $a = 50$, in only 16 of the 50 cases are there even added points at all.

3. We have also investigated other values of a; our simulation results indicate that the effectiveness of the algorithm depends on both b and r.

We next consider, though less completely, the case where the target consists of two squares, each with sides of length 0.1 parallel to the coordinate axes, whose centers are at fixed but unknown locations.

This case differs in several respects from that of one square. Most important, there is no characterization of the maximum likelihood estimator comparable to Theorem 6.1, and we have been unable to derive a finite algorithm for computation of it.

Numerical results for the case $a = 1000$, $b = 250$ are presented in Table 1.1, which contains for fifty simulated realizations the values of

Table 1.1 Simulation Results for "Two Squares"

Base rate $(b) = 250$ Added rate$(a) = 1000$

Bas	Add	1	2	1&2	LL	MLE 1	MLE 2	Act. Loc. 1	Act. Loc. 2
253.	21.	8.	13.	0.	1291.163	(9.00, 81.00)	(63.00, 81.00)	(5.37, 82.64)	(63.11, 81.37)
237.	20.	7.	13.	0.	1177.985	(27.00, 81.00)	(90.00, 9.00)	(42.61,85.74)	(87.51,10.68)
235.	19.	9.	10.	0.	1166.249	(0.00, 54.00)	(72.00, 45.00)	(76.66, 42.98)	(0.30, 52.53)
228.	19.	10.	9.	0.	1117.943	(27.00, 9.00)	(27.00, 27.00)	(11.44, 13.65)	(25.33, 28.08)
245.	20.	8.	12.	0.	1223.157	(9.00, 72.00)	(18.00, 72.00)	(21.49, 73.77)	(10.38, 70.64)
246.	24.	10.	14.	0.	1256.202	(36.00, 45.00)	(63.00, 81.00)	(61.02, 84.39)	(38.95, 46.19)
242.	17.	5.	12.	0.	1185.200	(45.00, 18.00)	(45.00, 27.00)	(27.81, 81.12)	(50.52, 21.53)
285.	15.	8.	7.	0.	1410.580	(0.00, 72.00)	(9.00, 0.00)	(71.09, 31.54)	(47.55, 2.28)
302.	19.	7.	12.	0.	1541.016	(27.00, 63.00)	(72.00, 9.00)	(67.47, 8.57)	(29.08, 63.00)
230.	18.	8.	10.	0.	1131.511	(81.00, 9.00)	(90.00, 90.00)	(75.35, 4.56)	(88.70, 88.60)
221.	15.	7.	8.	0.	1055.940	(81.00, 36.00)	(81.00, 36.00)	(76.31, 33.43)	(61.73, 22.96)
235.	22.	12.	10.	0.	1182.814	(9.00, 54.00)	(45.00, 36.00)	(45.72, 34.86)	(74.89, 1.65)
243.	16.	7.	9.	0.	1187.419	(63.00, 81.00)	(81.00, 54.00)	(49.03, 58.66)	(80.34, 56.83)
259.	17.	12.	5.	0.	1273.580	(9.00, 36.00)	(9.00, 36.00)	(32.16, 87.91)	(80.36, 28.54)
250.	13.	8.	5.	0.	1212.724	(81.00, 45.00)	(90.00, 18.00)	(85.34, 18.99)	(89.65, 65.02)
257.	18.	9.	9.	0.	1275.762	(0.00, 72.00)	(36.00, 36.00)	(42.77, 7.65)	(0.08, 73.80)
254.	22.	10.	12.	0.	1284.503	(63.00, 9.00)	(63.00, 63.00)	(43.52, 46.54)	(58.25, 64.57)
254.	21.	10.	11.	0.	1277.372	(18.00, 45.00)	(18.00, 81.00)	(67.32, 27.14)	(33.25, 57.02)
224.	25.	11.	10.	4.	1144.470	(81.00, 54.00)	(90.00, 54.00)	(80.59, 57.30)	(88.93, 56.75)
265.	22.	14.	8.	0.	1343.629	(63.00, 27.00)	(63.00, 45.00)	(64.07, 24.27)	(30.50, 5.92)
274.	13.	6.	7.	0.	1342.363	(9.00, 9.00)	(9.00, 9.00)	(39.14, 3.14)	(5.21, 8.53)
232.	15.	10.	5.	0.	1116.333	(63.00, 54.00)	(72.00, 9.00)	(69.08, 56.64)	(66.52, 13.15)
233.	22.	13.	9.	0.	1171.771	(9.00, 18.00)	(54.00, 54.00)	(55.67, 53.32)	(20.51, 49.95)
264.	14.	4.	10.	0.	1293.936	(0.00, 45.00)	(36.00, 9.00)	(58.64, 3.64)	(6.55, 47.84)
270.	18.	8.	10.	0.	1357.198	(18.00, 63.00)	(81.00, 54.00)	(82.42, 51.77)	(15.58, 58.20)
237.	16.	8.	8.	0.	1159.118	(27.00, 9.00)	(72.00, 90.00)	(63.19, 89.96)	(29.57, 10.91)
246.	24.	10.	14.	0.	1265.859	(36.00, 27.00)	(72.00, 72.00)	(73.26, 71.91)	(35.99, 28.02)
228.	14.	5.	9.	0.	1086.507	(72.00, 9.00)	(81.00, 9.00)	(28.93, 25.20)	(77.27, 9.17)
249.	23.	15.	8.	0.	1265.636	(0.00, 72.00)	(63.00, 27.00)	(62.14, 32.94)	(0.36, 70.29)
248.	16.	9.	7.	0.	1216.636	(45.00, 18.00)	(54.00, 54.00)	(43.58, 14.87)	(71.02, 20.59)
238.	16.	9.	7.	0.	1164.640	(9.00, 45.00)	(54.00, 18.00)	(8.61, 49.15)	(59.53, 50.04)
267.	21.	7.	14.	0.	1358.807	(63.00, 27.00)	(90.00, 63.00)	(84.16, 60.83)	(65.08, 26.14)
248.	28.	17.	11.	0.	1286.112	(36.00, 36.00)	(45.00, 18.00)	(43.59, 17.14)	(31.31, 29.39)
207.	18.	9.	9.	0.	1001.299	(0.00, 54.00)	(90.00, 54.00)	(9.63, 59.09)	(89.91, 55.89)
277.	17.	9.	8.	0.	1377.451	(18.00, 36.00)	(18.00, 54.00)	(22.84, 37.19)	(56.84, 47.83)
239.	15.	5.	10.	0.	1156.593	(36.00, 36.00)	(36.00, 90.00)	(60.16, 22.15)	(36.08, 40.71)
246.	24.	12.	12.	0.	1254.593	(36.00, 36.00)	(36.00, 63.00)	(30.07, 31.12)	(35.94, 64.69)
248.	19.	11.	8.	0.	1229.981	(0.00, 45.00)	(54.00, 36.00)	(3.90, 47.02)	(51.77, 40.48)
242.	24.	13.	11.	0.	1234.116	(18.00, 18.00)	(90.00, 36.00)	(89.85, 41.97)	(17.87, 20.52)
236.	9.	4.	5.	0.	1118.166	(9.00, 27.00)	(18.00, 0.00)	(8.94, 28.25)	(18.24, 0.70)
240.	20.	8.	12.	0.	1204.206	(18.00,27.00)	(81.00, 0.00)	(59.61, 27.57)	(18.94, 25.88)
258.	22.	12.	10.	0.	1314.636	(0.00, 27.00)	(72.00, 27.00)	(5.33, 22.38)	(71.77, 28.67)
249.	21.	9.	12.	0.	1255.593	(9.00, 81.00)	(9.00, 90.00)	(13.91, 67.23)	(6.74, 82.96)
259.	16.	12.	4.	0.	1270.934	(0.00, 0.00)	(0.00, 54.00)	(2.97, 58.99)	(47.19, 27.32)
261.	21.	9.	12.	0.	1319.241	(9.00, 54.00)	(36.00, 18.00)	(7.48, 55.08)	(21.13, 82.17)
257.	15.	8.	7.	0.	1257.588	(9.00, 36.00)	(54.00, 45.00)	(10.07, 30.94)	(31.63, 28.11)
258.	27.	8.	19.	0.	1338.415	(36.00, 0.00)	(36.00, 9.00)	(65.66, 38.28)	(39.44, 5.15)
244.	11.	4.	7.	0.	1162.114	(0.00, 90.00)	(27.00, 72.00)	(2.56, 86.42)	(28.55, 73.14)
253.	23.	10.	13.	0.	1287.112	(36.00, 54.00)	(45.00, 54.00)	(46.60, 87.46)	(38.87, 55.09)
250.	11.	8.	3.	0.	1201.071	(72.00, 18.00)	(72.00, 27.00)	(73.54, 26.35)	(54.97, 46.29)

the number of base points

the numbers of added points in each of the two squares and their intersection as well

the maximum value of the log-likelihood function and associated approximate maximum likelihood estimators over a grid of candidate square pairs, computed by considering as candidates squares centered at .09, .18, .27, .36, .45, .54, .63, .72, .81, and .90, and performing an exhaustive search

the actual locations of the two squares

The relative brightness $r = (a + b)/b$ is 5 in this case, and despite it being only approximate, the algorithm performs reasonably well: of the 50 cases,

In 16, both squares are located with reasonable accuracy.

In 28, one square (but not the other) is located.

In only 6 is neither square located, and these correspond in general to low numbers of added points.

6.3   Square or Disk?

Finally we consider the case that the target is, with probability .5, a square with sides of length 0.1 parallel to the axes and center with fixed but unknown location in $[.05, .95] \times [.05, .95]$, and, also with probability .5, a disk of radius 0.0564 and center having fixed but unknown location in $[.0564, .9436] \times [.0564, .9436]$. The radius of the disks is chosen so that they and the squares have the same area (0.01).

The objective is to determine the type (square or disk) and location of the target. Our analysis consists of

1. Calculation of maximum likelihood estimators under the hypotheses that the target is a square and a disk, respectively. These are effected using the algorithm described in Theorem 6.1 and a corresponding version for disks.
2. Calculation of the associated maximum values of the log-likelihood, which we denote by M_s and M_d for the square and disk, respectively.
3. Resolution of the object type by means of a *likelihood ratio test*: the object is deemed a square if $M_s > M_d$ and a disk if $M_d > M_s$. The type is undetermined if $M_s = M_d$.

The following table summarizes numerical properties of data obtained via Monte Carlo simulations. In this table, "correct" means that *either* the shape is identified correctly as a square or disk by the likelihood ratio test described above *or* that the likelihood ratio test is indeterminate in the sense that $M_s = M_d$ *and* that the shape is located within 0.05.

b	r	Cases	No. correct	% correct
1000	6.0	50	33	66.7
500	3.50	100	38	38.0
250	2.25	100	13	13.0
125	1.625	100	8	8.0
50	1.25	100	2	2.0

Key qualitative features of the data are:

1. The accuracy of the procedure degenerates rapidly as either the relative brightness r or the number of added points in the target decreases. For $r = 6$ the accuracy is quite good, whereas once $r = 2.25$ or less, there is effectively no accuracy.
2. The locations of the two maximum likelihood estimators are highly correlated, particularly for larger values of r, but even for values as small as 1.25. We have performed no formal statistical analysis; however, the following table shows the numbers of cases in which the locations of the two estimators differ by less than 0.05.

r	No. of "coincident" estimators
6.00	44/50 = 88%
3.50	79/100
2.25	67/100
1.625	54/100
1.25	55/100

3. In a number of cases, at least for higher values of r, in which the target type is determined incorrectly, nevertheless either the correct or the incorrect maximum likelihood estimator (or both) locates the target relatively accurately, as the following table shows.

r	Shape errors	"Correct" locations
6.00	10	8
3.50	17	6
2.25	15	1
1.625	10	0
1.25	11	0

4. Of shape errors, more are incorrect identifications of squares as disks than *vice versa*. We have no explanation of this.

For additional discussion of this model, see Karr (1988).

REFERENCES

Besag, J. (1974). Spatial interaction and the statistical analysis of lattice systems. *J. Royal Statist. Soc. B*, *36*, 192–236.

Besag, J. (1986). On the statistical analysis of dirty pictures. *J. Royal Statist. Soc. B*, *36*, 259–302.

Cross, G. R., and Jain, A. K. (1983). Markov random field texture models. *IEEE Trans. Patt. Anal. and Mach. Intell. 5*, 25–39.

Dempster, A. P., Laird, N. M., and Rubin, D. B., (1977). Maximum likelihood from incomplete data via the EM algorithm. *J. Roy. Statist. Soc. B*, *39*, 1–38.

Geman, D., and Geman, S. (1984) Stochastic relaxation, Gibbs distributions, and the Bayesian restoration of images. *IEEE Trans. Patt. Anal. and Mach. Intell.*, *6*, 721–741.

Geman, S., and Hwang, C. R. (1982). Nonparametric maximum likelihood estimation by the method of sieves. *Ann. Statist.*, *10*, 401–414.

Grenander, U. (1981). *Abstract Inference*. Wiley, New York.

Hajek, B. (1988). Cooling schedules for optimal annealing. *Math. Operations Res.*, *13*, 311–329.

Kak, A. C., and Rosenfeld, A. (1982). *Digital Picture Processing, I. (2nd ed.)*. Academic Press, New York.

Karr, A. F. (1986). *Point Processes and their Statistical Inference*. Marcel Dekker Inc., New York.

Karr, A. F. (1987). Maximum likelihood estimation for the multiplicative intensity model, via sieves. *Ann. Statist.*, *15*, 473–490.

Karr, A. F. (1988). Detection and identification of geometric features in the support of two-dimensional Poisson and Cox processes. *Technical report*, The Johns Hopkins University, Baltimore.

Karr, A. F., Snyder, D. L., Miller, M. I., and Appel, M. J. (1990). Estimation of the intensity functions of Poisson processes, with application to positron emission tomography. (To appear).

Kindermann, R., and Snell, J. L. (1980). *Markov Random Fields*. American Mathematical Society, Providence.

Lindsay, B. G. (1983a). The geometry of mixture likelihoods, I: A general theory. *Ann. Statist.*, *11*, 86–94.

Lindsay, B. G. (1983b). The geometry of mixture likelihoods, II: The exponential family. *Ann. Statist.*, *11*, 783–792.

Mandelbrot, B. (1983). *The Fractal Geometry of Nature*. W. H. Freeman, New York.

Matheron, G. (1975). *Random Sets and Integral Geometry*. Wiley, New York.

Ripley, B. (1988). *Statistical Inference for Spatial Processes*. Cambridge University Press, Cambridge.

Serra, J. (1982). *Image Analysis and Mathematical Morphology*. Academic Press, London.

Shepp, L. A., and Kruskal, J. B. (1978). Computerized tomography: the new medical x-ray technology. *Amer. Math. Monthly*, *85*, 420–439.

Snyder, D. L., Lewis, T. J., and Ter-Pogossian, M. M. (1981). A mathematical model for positron-emission tomography having time-of-flight measurements. *IEEE Trans. Nuclear Sci.*, *28*, 3575–3583.

Snyder, D. L., and Miller, M. I. (1985). The use of sieves to stabilize images produced by the EM algorithm for emission tomography. *IEEE Trans. Nuclear Sci.*, *32*, 3864–3872.

Vardi, Y., Shepp, L. A., and Kaufman, L. (1985). A statistical model for positron emission tomography. *J. Amer. Statist. Assoc.*, *80*, 8–20.

Wegman, E. J. and DePriest, D. J., eds. (1986). *Statistical Image Processing and Graphics*. Marcel Dekker, Inc., New York.

Wu, C. F. J. (1983). On the convergence properties of the EM algorithm. *Ann. Statist.*, *11*, 95–103.

2

Edge-Preserving Smoothing and the Assessment of Point Process Models for GATE Rainfall Fields

Colin R. Goodall* and Michael J. Phelan†
Princeton University, Princeton, New Jersey

A methodology developed for statistical analysis of rainfall fields is applied to weather-radar statistics on rainfall observed over the tropical Atlantic during the GATE experiment. First, the problem of robust estimation of the intensity of rainfall is addressed using a novel, resistant, edge-preserving technique for multivariate smoothing of structured data from spatial and spatial-temporal processes. It is sensitive to the mesoscale organization underlying tropical rainfall and highlights persistent patterns of temporal variation. Secondly, the problem of the assessment of point-process models for rainfall fields is addressed. A procedure, involving nonlinear least-squares, is implemented in the assessment of a family of LeCamian models.

1 INTRODUCTION

The investigation of rainfall fields entails the analysis of data from spatial processes. These data refer to the statistics of random phenom-

*Partially supported by Army Research Office Contracts DAAL03-88-K-0045 and DAAL03-86D-0001.
†Partially supported by National Science Foundation Grant DMS-8800346.

ena that distribute in space and evolve in time. The example considered here is the amount of rain falling from convective-cloud systems situated over the tropical region of the Atlantic Ocean. In this, the intertropical-convergence zone, the distribution of rainfall changes with time as thermodynamic conditions evolve. From a variety of instruments including weather radar, the activity of tropical rainfall over the Atlantic Ocean was observed in the Summer of 1974 during the *Global Atmospheric Research Program's Atlantic Tropical Experiment* (GATE). These data exhibit highly-structured patterns in the distributions of rainfall which is our goal to uncover, display, and summarize here.

The organization and structure of precipitative systems is well observed and documented in, for example, Houze and Hobbs (1982). Tropical systems, as do other convective systems, typically show mesoscale (10 km to 1000 km) organization involving distinctive formations of stratiform clouds and convective raincells. These formations range from isolated, small, cumulus clouds to large clusters of cumulonimbus clouds towering above a canopy of cirrus clouds. The canopy is predominantly a stratiform structure characterized by widely spread, more-or-less uniform precipitation. The process of convection, however, embeds the canopy with subregions, characterized by more intense, less uniform rainfall activity, by inducing the occurrence of convective raincells. These subregions occur in clusters and are associated with vertical updrafts and turbulence in the atmosphere. In the context of a midlatitude-convective system over continental North America their activity is observed for example in Zawadzki (1973), from statistics on rainfall drawn from weather radar, while in the context of tropical-convective systems, their activity is observed for example in Houze and Betts (1981) using a variety of instruments including weather radar and aircraft. The patterns of surface rainfall produced by these cloud systems reflect their underlying structure.

The present work develops a methodology for the statistical analysis of rainfall fields which is applied to data drawn from Phase I of GATE. The data comprise rates of rainfall prepared from weather radar as described in Hudlow et al. (1980). We address two problems of statistical inference. The first is to obtain a relatively smooth estimate of the intensity of rainfall that is sensitive to the often-detailed mesoscale organization of convective systems. Mesoscale organization informs a large class of probability models proposed in the literature for rainfall fields, including Le Cam (1961) and Gupta and Waymire (1987). They represent the intensity of rainfall by a random field obtained by a smoothing transformation of a point process. Thus the second problem we address here is to obtain an assessment of the capability of these models to describe the spatial evolution of rainfall fields, where we emphasize representations of the sample path of the intensity process. With this, we address a current challenge in this area posed, for example, in Cox and Isham (1988).

One of the principal aims of the GATE experiment was to provide data for an analysis of the water and energy budgets retained in the process of heat transfer in the tropics. The accuracy of these analyses hinges upon the accuracy of the observed rates of precipitation, which are generally viewed by meteorologists to be measured with error. As proposed in Hudlow and Patterson (1979), the effects of systematic bias due to large-scale fluctuations, such as atmospheric attenuation and positional shifts of the ships, are corrected for by Hudlow et al. (1980). Nevertheless, there remains error of measurements whose source extends beyond any adjustments aimed at attenuating systematic bias.

We resolve the problem of robust estimation of the intensity of rainfall here by a careful choice of a spatial smoothing technique. The success of the technique lies in its ability to preserve the mesoscale organization induced by the occurrence of convective raincells. Their occurrence is manifest in surface rainfall which has the appearance of convex subregions of relatively intense precipitative activity. Therefore, distributions of rainfall yield structured data, containing pronounced geometrical or other organizational features embedded in the spatial field. We choose to apply and extend a technique for resistant smoothing of multivariate data described in Goodall and Hansen (1988) called headbanging. The technique preserves edges and is sensitive to the geometrical structure underlying the data. Moreover, it is shown in Hansen (1989) to be effective relative to other techniques at preserving such structure.

Evolution of rainfall fields occur for example as the cloud system moves with the wind and dissipates with a storm's decay. Hence, in applying smoothing techniques to such data we face the problem of highlighting persistent patterns of temporal variation. We handle this problem here by extending from a spatial to a spatial-temporal technique of headbanging. Its implementation results in a family of spatially smoothed rainfall fields evolving in time; the spatial and temporal structure underlying the data is preserved, especially persistent features, and transient, spatially-isolated features eliminated.

Statistical models based on point processes are fundamental in the literature to phenomenological models for rainfall fields. Many of these draw upon the mesoscale features of precipitative systems, taking the convective raincell as the primitive for points. In such models, raincells are generated in clusters from a hierarchy of point processes in both space and time, whereby the resultant ensemble is supposed to mimic the observed geometry of mesoscale organization. The intensity of rainfall is represented by a smoothing transformation of the point process generating raincells. In effect, each raincell is endowed with a water content which is then spread in space and time by the chosen smoothing kernel.

Historically, the mathematical foundation for point-process models of precipitation was established by Le Cam (1961). For a range of precipitative systems, a general framework for operations on point processes involving subordination, clustering, and smoothing is rigorously developed. These ideas influence contemporary efforts to model rainfall as statisticians and hydrologists collaborate in formulating models of rainfall (Smith and Karr, 1983, 1985; Rodriguez-Iturbe et al., 1986; Rodriguez-Iturbe et al., 1987, 1988). These specific references include a healthy emphasis on the statistical analysis of actual rainfall as gathered from networks of *rain gauges*. While each is concerned with extratropical, continental rainfall, they are distinguished in their relative attention to spatial, temporal, or both aspects of its variation.

The probability models for rainfall in Waymire et al. (1984) and Gupta and Waymire (1987) are direct descendants of Le Cam's framework. Using the indicated operations on point processes, these models obtain the local rainfall by a smoothing transformation of a Poisson-cluster process of points generating raincell centers. The transformation involves a Gaussian and an exponential kernel for respectively spreading rainfall in space and in time. As an idealization of mesoscale organization, the geometry and kinematics of the model apparently describe the anatomy of tropical-convective systems. A statistical analysis of the model is given in Eagleson et al. (1987) and in Islam et al. (1988). These analyses assess large-sample predictions of spatial-temporal variation in rainfall from an air-mass thunderstorm in the southwest United States using statistics drawn from a network of rain gauges. Parameter estimators are obtained by the method of moments, and a number of favorable conclusions about the model are drawn. It remains to extend the analysis to the broader class of LeCamian models for rainfall fields, and to address the fundamental issue of whether such models are able to adequately *represent* rainfall fields.

Here we formulate a generalization of the model in Gupta and Waymire (1987) that retains the choice of geometry and kinematics as originally proposed. Leaving the exact stochastic mechanism for the process unspecified for now, we aim to assess this specific LeCamian representation of rainfall fields. Our approach is to fit the raincell-point process using nonlinear least-square and graphically assess the model's ability to represent actual rainfall fields. Note well that our emphasis is on representing sample intensity processes, rather than on large-sample statistical parameters. This analysis and our approach make a novel and, we believe and are informed, much needed contribution to the analysis of rainfall fields.

The paper proceeds in Section 2 with a description of the objectives of the GATE experiment including details of the data from Phase I of the experiment. A brief exposition of tropical convection provides the framework for understanding the structured distributions of rainfall observed in the GATE data. The edge-preservative headbanging smoother is applied to these data and its sensitivity to the mesoscale organization of raincells examined. In Section 3 we summarize the mathematical details of Le Cam's framework for models of precipitation. We focus on a particular model that is a generalization of the model developed in Gupta and Waymire (1987). In Section 4 we describe and implement a procedure for assessing the ability of such models to adequately represent rainfall fields. Finally, we address the implications of the methodology and directions for further research in the analysis of rainfall fields in short concluding remarks, in Section 5.

2 EDGE-PRESERVING SMOOTHING OF TROPICAL RAINFALL

2.1 Tropical Convection and GATE

Tropical convection is the principal means of heat transfer from the earth's surface to the troposphere, the lowest atmospheric layer. Maritime convection entails the release and vertical displacement of latent heat stored in the ocean during its heating by the sun. There is an obvious diurnal cycle to this process in that maritime convection occurs predominantly during the early morning hours. Over the tropical Atlantic, for example, this cycle is modulated by incoming waves of easterly wind.

Tropical convection manifests in a patchwork of cloud systems occurring in a narrow corridor at the center of the equatorial belt called the *intertropical-convergence zone* (ITCZ). The trade winds of both hemispheres converge in the ITCZ, amplifying updrafts of convection due to the release of surface heat. Mesoscale convection concentrates locally in groups of cumulonimbus clouds connected, in their mature and dissipative stages, by a common shield of cirrus cloud. The shield is essentially a canopy of stratiform cloud having horizontal dimension on the order of 100 to 1000 kilometers. Towering columns of cumulo-nimbus cloud erupt through the canopy under the amplified force of convective-heat release. These are the so-called convective raincells, having horizontal dimension on the order of 10 to 100 kilometers. The ensemble of cirrus and cumulonimbus cloud is called a cloud cluster, and its evolving patterns of surface-level precipitation is of particular interest here.

During the summer of 1974, the Global Atmospheric Research Program conducted the Atlantic Tropical Experiment, or GATE, in an effort to better understand tropical convection. A broad discussion of the historical, scientific, and statistical issues surrounding the study of convection in GATE is found in Houze and Betts (1981). Briefly, the scientific aims of the GATE experiment were to provide a description of the internal organization and structure of cloud clusters and to measure the transport of heat, moisture, and momentum of such systems. In relating these variables to large-scale movement of the tropical atmosphere, the goal was to further the basic understanding of tropical convection and its role in models of global-atmospheric circulation.

A practical outcome of the GATE experiment was the creation of an extensive data base of meteorological measurements over the study zone. These were made with the combined instrumentation of satellites, weather radar, aircraft, and upper-air balloons. Thus, the data include images of the large-scale development of cloud patterns and records of the three-dimensional field of precipitational intensity, as well as cloudmicrophysical variables. The rainfall data analyzed here is drawn from the weather-radar observations; this module of the experiment is now described in further detail. The radar observations were drawn from a network of C-band, digital radars stationed on board four ships. The center of the array of ships was about 8° North and 23° West, placing it on or about the ITCZ and about 700 kilometers southwest of Dakar, Senegal. The rainfall data analyzed here is from two of these ships; namely the Oceanographer and the Researcher. The Oceanographer was positioned about 8.5° North and 23.5° West, with the Researcher being positioned approximately 160 kilometers due south. Although data was being collected over the entire Summer of 1974, our analysis is confined to that collected during Phase I of GATE, spanning the 18 days from June 28 to July 16.

Throughout Phase I of GATE, each ship recorded a three-dimensional field of radar reflectivity at 15-minute intervals. Using current standards for conversion, these data were transformed to rates of precipitation falling in a circular region of the Atlantic, having a diameter of 400 kilometers. Details on the digital-radar systems and the conversional algorithms are found in Hudlow et al. (1980). Within the field of observation, the converted measurements have an areal resolution of 4 kilometers, yielding a grid of 4 × 4-kilometer pixels with rainfall rates. These rates are actually areal (but not temporal) averages of the precipitation falling within each subregion. Thus, for each sampling time, the data essentially consist of a 100 × 100 image containing the rate of precipitation falling in each 16 square kilometer pixel in the field of observation.

In addition to the 15-minute data, hourly aggregations of the reflectivity records were converted to rainfall rates by Hudlow and Patterson (1979). As an illustration, the hourly rates of precipitation generated by a typical cloud cluster is shown in a grey-scale reproduction of its image in Figure 2.1, where white denotes zero intensity and black

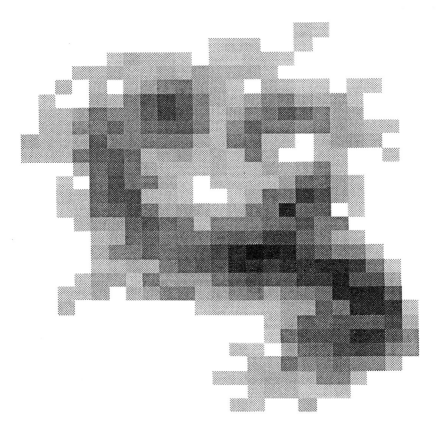

Figure 2.1 Cloud27.

denotes high intensity. This cloud cluster, called Cloud27, was isolated in the rainfall field observed during hour three of Julian Day 180. At the time, Cloud27 was in a mature stage, nearing its peak activity with rainfall rates within the shaded area ranging from about 0.3–0.7 mm/hr (millimeters per hour) to about 6–8 mm/hr. In Figure 2.1, as in all our images, north is up the page.

The distribution of rainfall from Cloud27 exhibits a highly-structured pattern, that we imagine manifests the underlying mesoscale organization of the system. In particular, note the S-shape ridge of high-intensity rainfall snaking through the image, which we imagine manifests the combined activity of a cluster of convective raincells. The two largest raincells of the cluster are located in the southeast corner of the image and an intermediate-sized raincell is located near the northwest corner of the image. In all, there appear to be six convective raincells present, each associated with a relative peak in rainfall activity and rather sharp gradients in such activity along the outward directionals from their centroids. We used a data-analytic definition of raincells based upon the empirical study of convection in precipitation in Houze (1981), yielding subregions of relatively intense rainfall in the range of 6–8 mm/hr having areal extent about 64-square kilometers. In contrast, note the region of more or less uniform, low-intensity rainfall enveloping the darker ridge. We imagine this manifests the activity of the stratiform-cloud layer supporting the raincells. The stratiform structure is contained within a 10,000-square kilometer subregion located southwest in the field of observation. Since the observed rainfall pattern is thus identified with the geophysical constructs of raincell and stratiform cloud, we resolutely hold an eye toward preserving this structure when smoothing these data.

2.2 Edge-Preserving Smoothing and Headbanging

A suitable choice of smoothing technique for tropical rainfall fields is a procedure called headbanging. Headbanging is a median-based filter for smoothing multivariate data described in Goodall and Hansen (1989) and designed to be sensitive to the presence of geometrical structure that may be underlying the data. Briefly, in smoothing a pixel value headbanging determines a median-based summary drawn from a neighborhood of collinear triples centered on the pixel, as shown in Exhibit 1. The ends of each triple are ordered and placed in a low and high set, respectively. The low and high set medians are called the high screen and low screen. The smooth is the median of the value at the pixel and the two screens. The result of 10 iterations of headbanging

A.

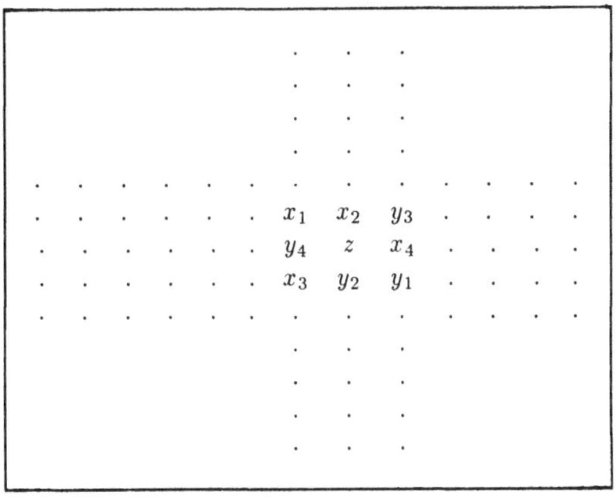

x_1 x_2 y_3
y_4 z x_4
x_3 y_2 y_1

B. Low set $= \{x_1 \quad x_2 \quad x_3 \quad x_4\}$
    High set $= \{y_1 \quad y_2 \quad y_3 \quad y_4\}$

C. L = low screen = median of low set
    H = high screen = median of high set

D. Smooth of z = median $\{L, z, H\}$

Exhibit 1 Spatial headbanging. (A) The neighborhood of z is defined by 4 triples (x_i, z, y_i), $i = 1, \ldots, 4$, labeled so that $x_i \leq y_i$ for each i. (B) The low end values and high end values are collected into low and high sets respectively. (C) The low and high set medians, called screens, are computed. (D) The smooth at the center position is the median of the two screens and the center observation z.

of Cloud27 is shown in Figure 2.2. Note that the actual procedure involves "twicing" in the manner usually recommended for median-based filters (Tukey, 1977).

The palette or grey scale used in Figure 2.2 is the same as that used in Figure 2.1, allowing for direct comparison. Now the rainfall rates within the shaded area range from about 0.18–0.6 mm/hr to about 4–5.7 mm/hr. Headbanging clearly produces a smoother representation of the precipitation generated by Cloud27, while preserving the mesoscale organization of its most prominent features. The smoothing effect of headbanging is particularly evident for example in the parts of the stratiform-cloud layer surrounding the raincells. The "hole" at center of Cloud27 in Figure 2.1 is rendered as a "valley" in the smoothed version. On the other hand, the spikes or narrow peaks that localize the incidence of convective activity have been leveled into broader-

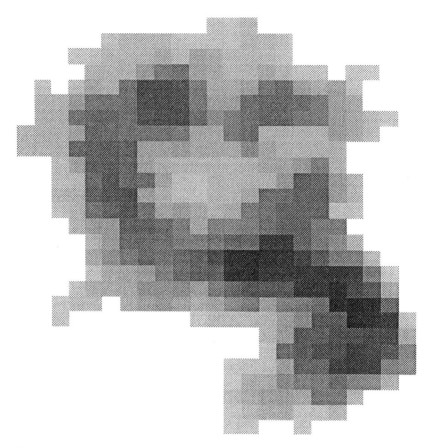

Figure 2.2 Spatial smooth of Cloud27.

based plateaus manifesting the effect of headbanging's resistance to extreme values. Notably, however, the integrity of the convective rain-cells as placed along the ridge of high-intensity rainfall discussed above is well preserved. And perhaps the most striking testimony to headbanging's ability to preserve edges is seen by examining the original and smoothed edge of the cloud system itself.

2.3 Spatial-Temporal Headbanging

An additional consideration in smoothing data from spatial processes is highlighting patterns of temporal variation. Recall that rainfall fields evolve over time through growth, dissipation, and movement of the generating cloud system in the prevailing winds. Not all features of a given cloud system at a given time will be well supported or persistent over adjacent times. Thus, it may be desirable to highlight those features that are supported in both space and time.

We extend the headbanging technique to obtain a resistant, edge-preserving procedure for smoothing simultaneously in space and time. In smoothing a pixel value the spatial version of headbanging uses a neighborhood of triples, each made up of the center pixel and nearby pixels, to determine high and low screens. With spatial-temporal headbanging (Exhibit 2) the neighborhood is extended to triples that include pixels in previous and later images that are near the center pixel. We separate out the purely spatial triples from the triples in space and time, and summarize their respective end values by a pair of low and high spatial screens, and a pair of low and high spatial-temporal screens. Instead of applying the spatial and spatial-temporal screens to the center value successively, we compute a median of 5, of the center value and the four screens, at each pixel. To smooth Cloud27 in space and time we include images one hour before (Cloud26, Figure 2.3) and one hour later (Cloud28, Figure 2.4). In general, the choice of width for the smoothing window over time is arbitrary. We have chosen a three-hour span since that is thought appropriate to the expected duration of persistent convective activity.

Each image in the GATE database is in the same position on the earth's surface. Thus the sequence of images could be visualized as a space $\times$ time data cube for the purpose of choosing nearby pixels. However, the appropriate neighborhood is not this geographically-fixed, Eulerian one, but a Lagrangian neighborhood that moves with the cloud system, in particular, with the raincell centers. The cloud system is moving predominantly in a southwesterly direction superimposed on which are movements of individual raincell relative to the ensemble. Indeed the reshaping of Cloud27 is apparent in Figures 2.3, 2.1, and 2.4, where in Figure 2.4 we see a more concentrated pattern than appears at earlier times. Thus, defining the neighbors to a pixel in Cloud27 from Cloud26 and Cloud28 requires that the rainfall images at hours two, three, and four be registered, using the convective raincell centers as landmarks and interpolating over their relative movements within the images. The registration map is described more fully below. Once the registration map is determined, headbanging incorporates the general algorithm for *irregularly* spaced data (Exhibit 2), described in Goodall and Hansen (1989), into the spatial-temporal context. The details are described in Goodall (1989).

In outline, first, the coordinates of the three images, Cloud26, Cloud27, and Cloud28, are fixed relative to the world coordinate sys-

A.　　　　　　Past　　　　　Present　　　　　Future

$$s_1 \quad t_2 \quad s_3 \qquad x_1 \quad x_2 \quad y_3 \qquad t_1 \quad s_2 \quad t_3$$
$$s_4 \quad s_5 \quad t_6 \qquad y_4 \quad z \quad x_4 \qquad t_4 \quad t_5 \quad s_6$$
$$s_7 \quad t_8 \quad t_9 \qquad x_3 \quad y_2 \quad y_1 \qquad t_7 \quad s_8 \quad s_9$$

B.　Low spatial set $= \{x_1 \ldots x_4\}$
　　High spatial set $= \{y_1 \ldots y_4\}$

　　Low spatial-temporal set $= \{s_1 \ldots s_9\}$
　　High spatial-temporal set $= \{t_1 \ldots t_9\}$

C.　Low (high) spatial screen $= L_s \ (H_s)$
　　Low (high) spatial-temporal screen $= L_{st} \ (H_{st})$

D.　Smooth of $z =$ median $\{L_{st}, L_s, z, H_s, H_{st}\}$

E.

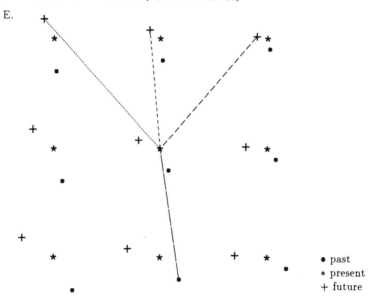

● past
* present
+ future

Exhibit 2 Spatial-temporal headbanging. (A) Without registration, the neighborhood of z contains 4 spatial triples, (x_i, z, y_i), $x_i \leq y_i$, $i = 1, \ldots, 4$, and 9 spatial-temporal triples, (s_i, z, t_i), $s_i \leq t_i$, $i = 1, \ldots, 9$. (B) Definition of low and high spatial and spatial-temporal sets. (C) The four screens. H_{st} for example is given by the median of the points in the high spatial-temporal set. (D) The smooth is the median of the four screens and the center observation z. (E) With registration, the spatial triples do not change but the spatial-temporal triples are now based on past (●) and future (+) pixel centers mapped into the present image. The mapping results in the irregularly spaced data shown in the figure. The set of approximately collinear triples that include the past point at lower-center is marked. Further details on the choice of triples for irregularly spaced data are provided by Goodall and Hansen (1988).

Figure 2.3 Cloud26.

tem. Then each of Cloud26 and Cloud28 is registered with Cloud27, yielding a nonlinear transformation of their respective coordinates to the coordinates of Cloud27. The mapping yields deformed ensembles of rainfall rates from Cloud26 and Cloud28 respectively at irregularly-spaced coordinate locations in the co-ordinate frame of Cloud27. (Actually we obtain deformed quadrilaterals, but with the median technique using the center location of each alone suffices.) The spatial screens of headbanging are constructed in the usual way from Cloud27 alone, whereas the temporal screens are constructed from triples with an end point in each of Cloud26 and Cloud28, drawn from among the nearest neighbors after registration. The spatial-temporal smooth of Cloud27

Figure 2.4   Cloud28.

is shown in Figure 2.5, using the same palette as used in Figure 2.1 for
Cloud27.

As expected, spatial-temporal headbanging of Cloud27 yields a
heavier smooth for Cloud27 than does spatial headbanging. In particular, many of the irregular features at the edge of the cloud system
are shaven away. Within this effect, there is a very modest extension
of the area of stratiform precipitation. Notwithstanding, the most persistent feature of the cloud system is both highlighted and well preserved,
that is, the broadly-spread region of low-intensity, more or less uniform
precipitation supporting an S-shaped ridge of high-intensity precipitation. This feature is shown to be well supported in time (cf. Figures
2.3, 2.1, and 2.4). In contrast, note the apparent attenuation of intense

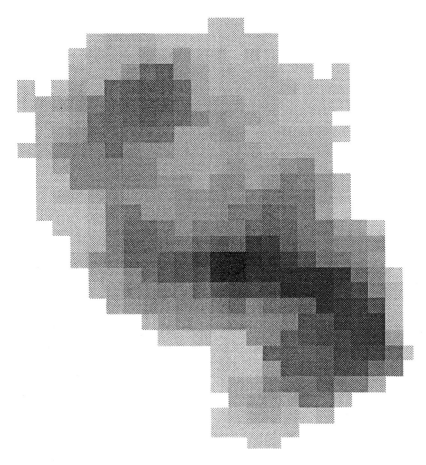

Figure 2.5 Spatial-temporal smooth of Cloud27.

precipitation in the northeast corner of the cloud. This is a subregion of convective activity in Cloud27 that is preserved in the spatial smooth (see Figure 2.2). Although a case may be made for such an effect, particularly for short-lived raincells we may also wish to twice spatial-temporal headbanging as a means to protect against potential over-smoothing.

2.4   Registration Map

We conclude this section on smoothing with a brief description of the registration map. The procedure treats the centers of the convective raincells as landmarks and registers them exactly. That is, the pixel location of each raincell is registered between images by a direct translation. This yields the vertical and horizontal displacements needed to map the existing raincells from, for example, Cloud26 to Cloud27. The remaining pixels that comprise the image of Cloud26 are translated using vertical and horizontal displacements obtained by interpolating over the displacements determined by the raincells. The interpolation scheme involves an adaptation of Akima's algorithm (Akima, 1978) as described in Goodall (1989). A graphical representation of the registration map is shown in Figure 2.6.

We focus first on the top two panels in Figure 2.6, showing the

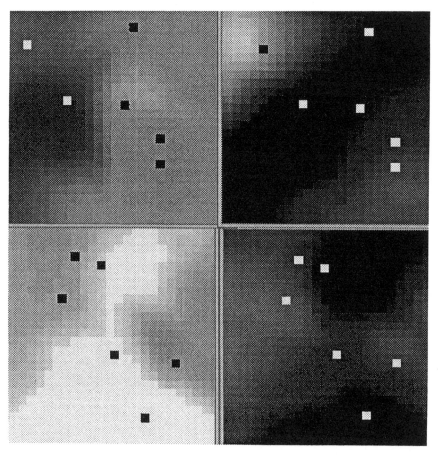

Figure 2.6 Registration map for Cloud27.

registration of Cloud26 with Cloud27. The planar coordinates are those of the imaging space of the clouds. The highlighted pixels denote the locations of the convective raincells as identified here; their color, black or white, being chosen to differentiate their location against the background. The top two panels code the translation by greyscale intensity in world coordinates, used to map pixel locations from Cloud26 to Cloud27; the left panel indicates vertical displacement and the right panel indicates horizontal displacement. Grey denotes no translation (same position) in both panels, lighter shades denote downward translation and darker shades denote upward translation in the left panel, while lighter shades denote eastward translation and darker shades denote westward translation in the right panel. Together, these panels show movement predominantly in the southwest direction from Cloud26 to Cloud27, but with a slight rotation in the clockwise direction. This translation is consistent with our graphical tracking of the cloud system between hours two and three, comparing Figures 2.1 and 2.3. The bottom two panels in Figure 2.6 show the analogous display of the registration map for mapping Cloud28 to Cloud27. Note that the interpretation of the greyscale is reversed in the lower panels where we are mapping the cloud backward in time; that is, the dark in both right-hand panels shows that the westward translation persists across both time intervals.

3 STATISTICAL MODELS FOR RAINFALL FIELDS

Statistical models based on point processes are fundamental to phenomenological models for rainfall fields. We describe here a statistical model based upon the mesoscale organization of precipitative systems as idealized in the mathematical framework in Le Cam (1961), that represents the surface intensity of rainfall by a smoothing transformation of a point process generating convective raincells. The model generalizes the one in Waymire et al. (1984), and in Gupta and Waymire (1987). We emphasize here the geometry and kinematics of these rainfall representations, rather than their probabilistic details.

3.1 Point Process of Convective Raincells

Let T denote a subset of $\mathbb{R}$ which is a time index set. Let S and V denote subsets of $\mathbb{R}^2$, representing respectively a region on the surface of the earth and a velocity index set. Let W and A denote subsets of $\mathbb{R}_+$ which are index sets for the water content of a raincell and its aging rate, respectively, and let Σ denote a subset of the set of symmetric, positive-definite matrices of order two, an index set for dispersion. Finally, let $(E, \mathbf{E})$ denote the product space $T \times S \times W \times V \times \Sigma \times A$ equipped with its Borel sets.

We introduce a random counting measure X on $(E, \mathbf{E})$ defined on a probability space $(\Omega, \mathbf{H}, \mathbb{P})$ to model the placement of convective raincells in time and space together with their water content, aging rate, velocity, and dispersion. Thus, $X(B)$ is a random variable counting the number of raincells whose characteristics lie in B, for B a Borel subset of E.

3.2 Rainfall Intensity

For each $(t, x) \in T \times S$, let $k(t, x, \cdot)$ denote a positive Borel function defined on E. Then

$$R(t, x) = \int_E X(db)k(t, x, b) \qquad ((t, x) \in T \times S) \qquad (3.1)$$

defines a random field over time and space, a space-time process for the evolving intensity of rainfall. Thus, for each $t \in T$, $(R(t, x), x \in S)$ denotes the distribution of rain falling over a region of the earth at time t, that evolves as the precipitative system changes with time.

According to equation (3.1), the intensity of rainfall is obtained by a smoothing transformation of the point process X. Although its choice is somewhat arbitrary, the role of the smoothing kernel k is to spread the water content of a raincell in space and time. This goal is achieved here as follows. For every $t, u \in T$, $x, y \in S$, $w \in W$, $v \in V$, $a \in A$, and $\sigma \in \Sigma$, we write $b = (u, y, w, v, \sigma, a)$ and assume that k satisfies

$$k(t, x, b) = w \exp\{-a|t - u|\} \exp\{-\|\sigma^{-1/2}(x - y - v(t - u))\|^2\}/2\pi|\sigma|^{1/2}$$
$$(3.2)$$

where $\|\cdot\|$ denotes the Euclidean distance in $\mathbb{R}^2$. This choice of kernel spreads water content by a *Gaussian-spreading* in space coupled to an *exponential-spreading* in time. Thus a raincell is both a point in the product space E and an equivalent, by (3.1, 3.2), continuum of Gaussian surfaces.

3.3 Geometry and Kinematics

The representation of rainfall fields according to (3.1) and (3.2) entails a particular choice of *geometry* and *kinematics* to underlie the precipitative system. A realization of X determines the initial placement of raincells in time and space with their respective water content, aging rate, velocity, and scale. Then the smoothing kernel spreads the water contents over space and time. The former is achieved by constructing a Gaussian surface centered *spatially* at each raincell; the amount of water spread to any point of S decays exponentially as the square of its scaled distance from the raincell's center. A raincell with $b = (u, y, w, v, \sigma, a)$ moves with velocity v and has location $y + (t - u)v$ at time t. This motion is applied in equation (3.2), so that each Gaussian surface tracks the same trajectory across S as its associated raincell. The raincell ages at rate a, where absolute age is measured relative to a time of maturation u. Thus, temporal spreading of the total water content $2w/a$ follows a doubly-exponential curve centered *temporally* at each raincell; the amount of water spread to any point of T decays exponentially as its absolute age scaled by its aging rate, allowing for growth and dissipation of a raincell. Integrating over space the total intensity of a raincell at time t is $w \exp(-a|t - u|)$.

The intensity of rainfall, as modeled above, is the sum of the intensities of the existing convective raincells. The model proposed in Waymire et al. (1984) has been modified in two ways. One is that the present model allows the water content, aging rate, velocity, and scale of a raincell to be random variables rather than a constant over all raincells. The second is that the present model allows the life cycle of each raincell to contain both a formative and a dissipative phase rather than only a dissipative phase. Indeed, the analysis of the previous section showed both phases were present. It also showed that the configuration of coverage by rainfall was often a patchwork of alternating "wet" and "dry" regions due to zero rainfall beyond the perimeter of the stratiform canopies. In contrast, the model allows each raincell to have infinite extent in all directions. Nevertheless for practical purposes the imposed exponential decay with distance is rapid, and no thresholding is required.

LeCamian Models

It remains to specify the random mechanism generating the points of X. In the LeCamian framework, these points are generated from a hierarchy of point processes. For representing large-scale precipitative systems, storm centers are generated first by an initial Poisson process having a specified mean measure. This realization is then transformed by a smoothing operation, obtaining a random measure to be used as the mean measure of a subordinated Poisson process in the next stage of generating points. In effect, the smoothing transformation endows each storm with magnitude and extent, while the second stage of generating points clusters squall systems, for example, about the eye of the storm. The algorithm is iterated until the mesoscale organization of convective raincells is attained. The raincells may be distinguished by their water content or other desired characteristic. Finally, note that the operations on point processes described above can be initialized at any level of mesoscale organization. In fact, the model proposed in Waymire et al. (1984) begins by generating "clusterpotentials" as centers for clusters of convective raincells. As in the general case, each stage generates points by a Poisson process, and the intensity of rainfall is represented by a smoothing transformation of the process generating raincells.

The primary interest here is in the geometry and kinematics of the LeCamian representation of rainfall fields. In Section 4 we apply a procedure designed to assess the representation treated here, which requires an estimate of the point field X. These estimates are obtained independently of the exact mechanism generating X, so that the procedure is robust to such misspecifications of the model while it is sensitive to the adequacy of the intended representation of rainfall fields.

4 MODEL ASSESSMENT VIA GATE RAINFALL

We assess the statistical model formulated in Section 3 for its ability to represent the spatial evolution of a typical, tropical cloud system, as observed during the GATE experiment. More specifically, given the point field generating a particular cloud cluster, we aim to assess the underlying geometry and kinematics imposed by the choice of the smoothing kernel shown above.

4.1 GATE Rainfall Data and Nonlinear Least-Squares

Let S denote the 400×400-kilometer square region over the surface of the Atlantic Ocean that contains the circular field of observation in the GATE experiment inscribed within. Let S_{ij} denote the pixel in the ith column and jth row of S, $i, j \in I = \{k: 0 \le k \le 99\}$, with S_{00} in the NW corner. Recall that the areal resolution of the digital-radar systems used during GATE is 4 kilometers.

Fix $i, j \in I$, and for $t \in T$ and $S_{ij} \subset S$, let $\bar{R}_t(i, j)$ denote the random variable given by

$$R_t(i, j) = \int_{S_{ij}} dx R(t, x) \qquad (4.1)$$

where $R(t, x)$ is given by equation (3.1). The $\bar{R}_t(i, j)$ involve the integral of rainfall intensity over S_{ij} summed over several raincells. Next, let $Y_t(i, j)$ denote the random variable given by the intensity of rainfall observed at hour t over S_{ij}. Our *statistical* model is

$$Y_t(i, j) = \bar{R}_t(i, j) + \epsilon_t(i, j) \tag{4.2}$$

where the $\epsilon_t(i, j)$ are independent measurement errors satisfying

$$E(\epsilon_t(i, j) \mid X) = 0 \quad \text{and} \quad E(\epsilon_t^2(i, j) \mid X) = \sigma^2 > 0 \tag{4.3}$$

where X denotes the point field that determines the $R(t, x)$. Thus

$$E(Y_t(i, j) \mid X) = \bar{R}_t(i, j) \tag{4.4}$$

and equations (4.2) and (4.3) specify a regression model for the $Y_t(i, j)$ having "parameter" X. We estimate the conditional mean shown in (4.4) using nonlinear least-squares, thereby assessing the hypothesized representation of the rainfall field conditional on X and without a full specification of the probability model. Although a fully specified probability model for X is the ultimate aim of this research, such a formulation is not included in the present analysis. However, given the number of raincells at every hour, these estimates can be put on a Bayesian footing using independent uninformative priors.

Detailed analysis of the estimates of X is performed on the cloud system introduced in Section 2. We monitor the system over a five-hour period, covering 0100 to 0500 hours of Julian day 180. For each $t = 1, \ldots, 5$, this yields rainfall rates $(Y_t(i, j), i, j \in I)$ over the field of observation at hour t. The problem is to estimate X, the raincell's characteristics, from the $Y_t(i, j)$, also yielding a smooth $R_t(i, j)$ of the rainfall field. Recall that the characteristics are constant over time (3.1, 3.2). Nevertheless we choose to fit a more general model in which we decouple the images in time. We remove the explicitly exponential-spreading and velocity components of the model, parameters (a, u, v), and allow *separate* parameters (w_t, σ_t, y_t), say, for each raincell at each time. Here X_t denotes the raincell ensemble at hour t. Note that we still identify raincells at successive times as the same cell, which permits us to assess the stationarity of the raincell's characteristics and to better understand the spatial evolution of the system in the context of the model and its adequacy.

For each $t = 1, \ldots, 5$, the nonlinear least-squares estimate $\hat{X}_t$ is obtained by Gauss–Newton optimization. The full details are to appear elsewhere. For each image and raincell in that image, the estimator yields current intensity (w_t), spatial location (y_t), and dispersion (σ_t). Their initial values were set respectively to the average intensity in the image per raincell, the inferred raincell centers as described in Section

2, and spherically symmetric dispersion with standard deviation 1 pixel. After fitting, the estimates comprise a time series of raincell character- istics over the five hour period.

4.2 Results

The least-squares fit for Cloud27 is shown in Figure 2.7. It represents the data (Figure 2.1), and especially the smoothed data (Figure 2.2), very well. Contours of the components of $\bar{R}_r(i, j)$ for the separate raincells are shown in Figure 2.8 for each hour. These quantile contours

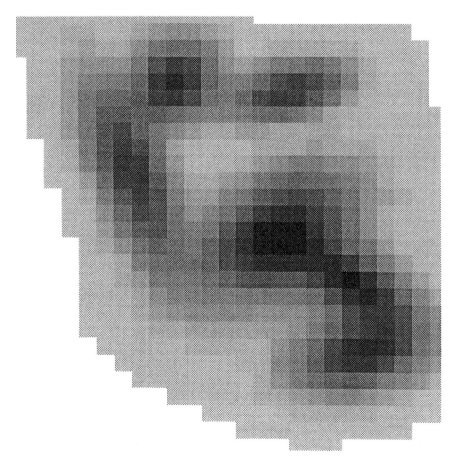

Figure 2.7 Nonlinear least squares fit of Le Camian model with Gausian kernels to Cloud27.

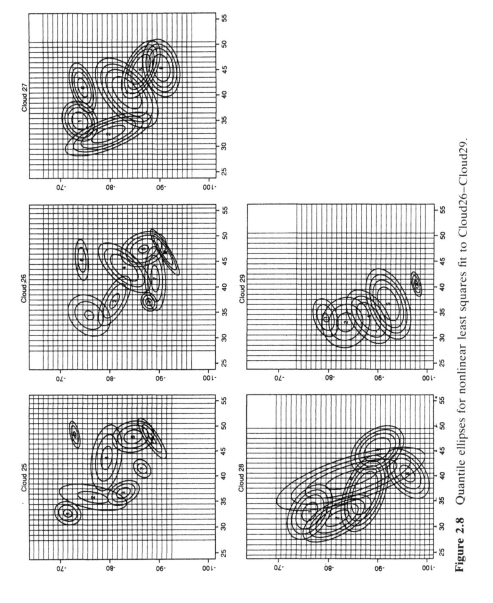

Figure 2.8 Quantile ellipses for nonlinear least squares fit to Cloud26–Cloud29.

are (outermost-in) at 0.30, 0.75, 1.5, and 3.0 mm/hr rainfall intensity. The innermost contours do not appear for all cells. The five panels of Figure 2.8 show the NW–SE S-shaped band of high-intensity rainfall, which is strongest in Cloud28, together with "satellite" cells, for example 3, 6 and 7 in Cloud25 and Cloud26. The raincells show a broadly-consistent pattern of growth and contraction. For example, cell 1 grows during the first hour, then intensifies during the second hour, elongates to Cloud28, then dissipates. Figure 2.9 shows the percentile ellipses, delimiting 50%, 80%, and 95% of the total rainfall intensity. These help localize some of the regions of rainfall activity. For example, the contraction of the percentile ellipses for cell 1 between Cloud26 and Cloud27, but not of the quantile ellipses, reflects the intensification near the raincell center.

Time series of six summary statistics are shown in Figure 2.10. Let the spectrum of σ be (p, q), $p \geq q \geq 0$ and θ be the orientation of the principal (p) axis to the horizontal WE direction. The six panels show raincell intensity, raincell areal scale $\sqrt{pq}$, major axis p, minor axis q, angular orientation θ, and anisotropy p/q. Note that $\sqrt{pq}$, p and q parametrize the shape and extent of the percentiles of the intensity distribution, as in Figure 2.9.

Not all raincells persist through the 5 hours. Notwithstanding, as may be further verified from Figure 2.8 and Figure 2.9, the time series are remarkably consistent. For example for cell 4 the intensity peaks at Cloud28, the areal scale, major axis and minor axis are almost constant around 2.25 pixels (9 km), 3.5 and 1.5 pixels respectively, the raincell anisotropy decreases gradually, and only the angular orientation oscillates across a 60° arc.

The estimates for cell 6 are consistent for Cloud25, Cloud26, and Cloud27 but very different for Cloud28. Inspection of the image for Cloud28, Figure 2.4, shows that cell 6 does persist in that image; however it is small and cell 6 of the Gauss-Newton solution is positioned elsewhere, and has the role of an "omnibus" layer spanning several cells. Clearly the atypical orientation of cell 4 in Cloud28 is explained by the prominence of cell 6. Cell 2 has an analogous role, as omnibus layer, in Cloud25. The atypical orientation of cell 4 in Cloud25 is associated with the atypical orientation of cell 2 there, that is, in turn, associated with the position of cell 3. Notice that cell 2 is to the other side of cell 3 in Cloud26.

The analysis presented here and in Figures 2.8–2.10 is not a final one. It does graphically highlight the dependencies between the raincells. The results are in fact surprisingly consistent across time. Further

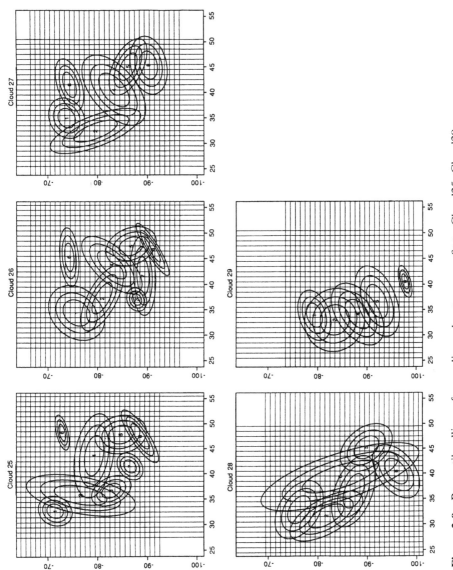

Figure 2.9 Percentile ellipses for nonlinear least squares fit to Cloud25–Cloud29.

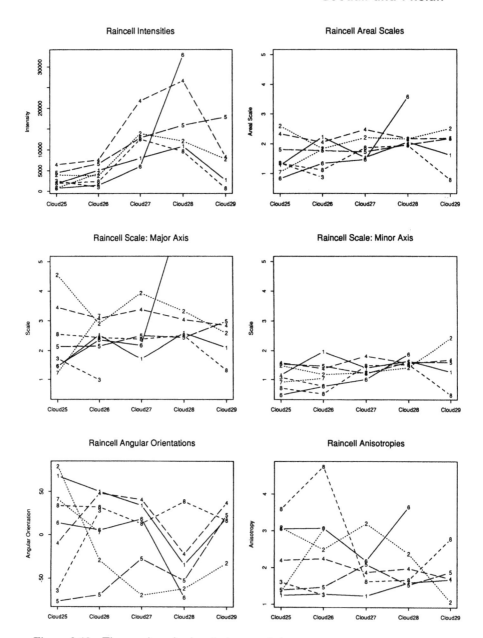

Figure 2.10   Time series of raincell characteristics.

analysis, not presented here, includes elimination of cell 6 at Cloud28 and possibly cell 3 at Cloud25 and Cloud26. These reductions help make the Gauss-Newton solutions compatible with our LeCamian model, (3.1, 3.2). Note further that the parameter estimates in the decoupled model are highly correlated. We are led to fit the original coupled model to enforce compatibility with (3.1, 3.2), and improve the conditioning of the system.

As an alternative, in the spirit of model assessment, consider the contours of Figure 2.8 and Figure 2.9. There is considerable overlap among the cells, and some very extensive cells (cell 2 in Cloud25, cell 6 in Cloud28). Clearly even a small cloud system can not be represented by a small number (5–8) of substantially *non*overlapping Gaussian kernels. At the minimum the model must be supplemented by a separate term for the stratiform layer. This improves the conditioning of the model and the separation of cells.

4.3 Recoupled Equations: Aging Rate and Water Content

For a given raincell, let I_t denote the intensity at hour t, where according to equation (3.2), I_t is given by

$$I_t = W \exp\{-A|t - T_0|\} \qquad (4.5)$$

where W, A, and T_0 are random variables respectively denoting water content, aging rate, and time of maturation of the raincell. The raincell intensities in Figure 2.10 estimate the I_t. The quantity $|t - T_0|$ denotes age, and according to (4.5), we model intensity during both the formative and dissipative stage in the life cycle of a raincell. We readily identify T_0 from the intensities in Figure 2.10, and on passing to the logarithmic scale, we fit W and A using ordinary least-squares. The results are displayed in Table 2.1.

The aging rates in Table 2.1 give the exponential *rate of growth and dissipation per hour* of the raincells with age. These show raincells one, two, four, and five to age at similar rates, being roughly 40% the rate of aging of raincells six and eight. The residuals from these fits suggest the need to separately fit a rate of growth and a rate of dissipation. Here, unfortunately, that analysis entails fitting many of the dissipation rates from only two points; a problem that affects the aging rates of raincells three and seven in Table 2.1. Moreover, we tentatively argue from the data that intensity may vary as some power of age, suggesting an interesting direction for further research.

The water contents in Table 2.1 are *proportional to millimeters* of rainfall in a 16-square kilometer area, where by far the most massive

Table 2.1 Aging Rate and Water Content

Raincell	Aging rate	Water content (2 W/A)
1	0.50	376
2	0.42	567
3	0.94	53
4	0.48	963
5	0.36	1167
6	1.29	406
7	1.71	51
8	1.22	234
Total:	–	3817
Mean:	0.86	477
Standard Deviation:	0.50	406

raincells are four, five, and six. To obtain a measure of its density per unit area at a given hour, we scale the water content of a raincell to its areal scale, and we find that raincell five is dense relative to raincell four up to about hours four and five.

4.4 Kinematics

According to the model, the time series of raincell locations estimate the initial position of the raincell translated in the direction of its velocity. That is, for a given raincell, let μ_t denote its planar location in world coordinates at hour t. According to equation (3.2), μ_t is given by

$$\mu_t = \mu_0 + (t - T_0)v \qquad (4.6)$$

Table 2.2 Raincell Velocities

Raincell	Horizontal Velocity	Vertical Velocity	Angular Orientation
1	0.07	−1.62	87.6
2	−1.14	−1.45	127.9
3	0.35	−5.08	85.9
4	−2.63	−2.19	140.5
5	−2.47	−1.77	144.5
6	−3.06	−3.51	131.3
7	0.15	−2.57	86.5
8	−2.03	−2.31	131.3
Mean:	−1.35	−2.56	120.7

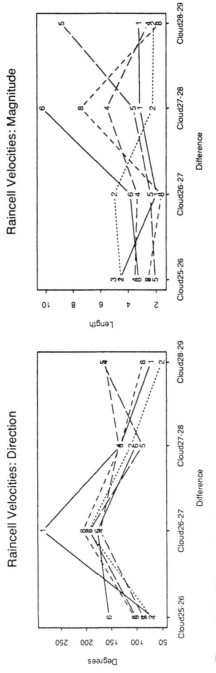

Figure 2.11 Time series of raincell velocities.

where μ_0 denotes initial position, v denotes velocity, and T_0 is defined at (4.5). We fit μ_0 and v from the estimated μ_ts using ordinary least-squares. The results are displayed in Table 2.2.

The velocities in Table 2.2 give the horizontal and vertical velocity in units of *four kilometers per hour* for each raincell. These indicate that many of the raincells are moving horizontally and vertically as much as 8 or 9 kilometers in an hour. Table 2.2 also shows the orientation of velocity to the horizontal eastwest direction. These are consistent with the prevailing wind direction. Alternatively, we consider the time series of first-difference velocities taken over the hours. The results are displayed in polar coordinates in Figure 2.11. Between each hour, the raincells move broadly in the southerly or the southwesterly direction, but the raincells do not appear to move lockstep. The magnitude are in accordance with expectation, showing movement between eight and eighteen kilometers per hour, with some exceptions occuring between hours three and four.

5 CONCLUDING REMARKS

We analyzed the spatial evolution of rainfall from a mesoscale precipitating feature called a cloud cluster observed during Phase I of GATE. We emphasized the mesoscale organization of the system, featuring the geophysical constructs of stratiform cloud and convective raincell. We implemented an edge-preserving smoothing technique called headbanging, and a procedure for assessing point process models for rainfall fields.

Headbanging delivers smoothed versions of rainfall fields that are true to the mesoscale organization of the generating cloud system. We emphasize its ability to preserve edges, while highlighting persistent patterns of convective and stratiform rainfall activity. A spatial-temporal version of headbanging is introduced here for the first time. The technique appears to have the same promise as headbanging in this context, but its properties need to be further explored.

We believe that the LeCamian framework for models of precipitation merits vigorous attention. We followed this belief in assessing the geometry and kinematics of a simple model within the general framework. We find that these may be more complex than initially supposed. We do find that the Gaussian geometry has some merit, but this depends critically on the age of the system, and suggests the need for a proper lifetime analysis of raincells and their characteristics.

REFERENCES

Akima, H. (1978). A method of bivariate interpolation and smooth surface fitting for irregularly distributed data points. *ACM Transactions on Mathematical Software*, *4*, 148–159.

Cox, D. R. and Isham, V. (1988). A simple spatial-temporal model of rainfall, *Proc. R. Soc. Lond.*, *A915*, 317–328.

Eagleson, P. S., Fennessey, N. M., Qinliang, W., and Rodriguez-Iturbe, I. (1987). Application of spatial poisson models to air mass thunderstorm rainfall. *J. Geophys. Res.*, *92*, 9661–9678.

Goodall, C. R. (1989). Nonlinear filters for ATR/laser radar image enhancement. Report for U.S. Army Center for Night Vision and Electro-Optics, 152pp.

Goodall, C. R. and Hansen, K. (1988). Resistant smoothing of high-dimensional data, *Proceedings of the Statistical Graphics Section*, American Statistical Association.

Gupta, V. K., and Waymire, E. (1987). On Taylor's hypothesis and dissipation in rainfall, *J. Geophys. Res.*, *92*, 9657–9660.

Hansen, K. (1989). Ph.D. Thesis, Princeton University, Princeton, New Jersey.

Houze, R. A. (1981). Structure of atmospheric precipitation systems: a global survey, *Radio Science*, *16*, 671–689.

Houze, R. A. and Betts, A. K. (1981). Convection in GATE, *Rev. Geophys. Space Phys.*, *19*, 541–576.

Houze, R. A., and Hobbs, P. V. (1982). Organization and structure of precipitating cloud systems, *Adv. Geophys.*, *24*, 225–315.

Hudlow, M. D. and Paterson, P. (1979). GATE Radar Rainfall Atlas, NOAA Special Report.

Hudlow, M. D., Patterson, P., Pytlowany, P., Richards, F., and Geotis, S. (1980). Calibration and intercomparison of the GATE C-band weather radars, *Technical Report EDIS 31*, 98 pp., Natl. Oceanic and Atmos. Admin., Rockville, Maryland.

Islam, S., Bras, R. L., and Rodriguez-Iturbe, I. (1988). Multidimensional modeling of cumulative rainfall parameter estimation and model adequacy through a continuum of scales, *Water Resource Res.*, *24*, 985–992.

Le Cam, L. M. (1961). A stochastic description of precipitation, in *Proceeding of the Fourth Berkeley Symposium on Mathematical Statistics and Probability*, J. Neyman, (ed.), *3*, 165–186, Berkeley, California.

Rodriguez-Iturbe, I., Cox, D. R. and Eagleson, P. S. (1986). Spatial modelling of total storm rainfall, *Proc. R. Soc. Lond. A*, *403*, 27–50.

Rodriguez-Iturbe, I., Cox, D. R., and Isham, V. (1987). Some models for rainfall based on stochastic point processes, *Proc. R. Soc. Lond. A*, *410*, 269–288.

Rodriguez-Iturbe, I., Cox, D. R., and Isham, V. (1988). A point process model for rainfall: further developments, *Proc. R. Soc. Lond. A.*, 417, 283–298.

Smith, J. A. and Karr, A. F. (1983). A point model of summer season rainfall occurrences, *Water Resource Res.*, *19*, 95–103.

Smith, J. A. and Karr, A. F. (1985). Parameter estimation for a model of space-time rainfall, *Water Resource Res.*, *21*, 1251–1257.

Tukey, J. W. (1977). *Exploratory Data Analysis*, Addison-Wesley, Reading, MA.

Waymire, E., Gupta, V. K., and Rodriguez-Iturbe, I. (1984). Spectral theory of rainfall intensity at the meso-$beta$ scale, *Water Resource Res.*, *20*, 1453–1465.

Zawadzki, I. I. (1973). Statistical properties of precipitation patterns, *J. Appl. Meteorol.*, *12*, 459–472.

3
Likelihood Methods for Diffusions with Jumps

Michael Sørensen
Aarhus University, Aarhus, Denmark

Likelihood methods are developed for inference about parameters that determine the drift and the jump mechanism of a diffusion with jumps. The data are supposed to be a continuously observed sample path. The likelihood function is derived, and the score function is shown to be a martingale. From this property the asymptotic distribution of the maximum likelihood estimators and of some test statistics is derived. Various types of Fisher information are considered. It is demonstrated how the theory simplifies and can be sharpened when a part of the likelihood function has the form of an exponential family. Four examples are discussed.

1 INTRODUCTION

The powerful tools of stochastic calculus are finding their way into many branches of applied probability and statistics these years. These methods have enabled analysis of more complicated models than could be handled earlier. An aspect of this development is the growing use of diffusions with jumps. A simple example of a process of this type is the solution of the stochastic differential equation

$$dX_t = \mu(X_t)\, dt + \sigma(X_t)\, dW_t + \delta(X_t)\, dZ_t$$

where $X_0 = x_0$, W_t is a Wiener process and Z_t is a compound Poisson process. Recent applications of diffusions with jumps are the general stock price model studied by Aase (1986a,b) and the stochastic differential equation approach to soil moisture taken by Mtundu and Koch (1987). For other applications in hydrology, see Bodo, Thompson, and Unny (1987). A population model in terms of a diffusion with jumps is studied in Section 5.

The aim of this chapter is to investigate likelihood methods for inference about parameters that determine the drift coefficient and the jump mechanism. The data are assumed to be a continuously observed sample path. From data of this kind it is impossible to apply likelihood methods to inference about the diffusion coefficient. We shall, however, give an estimator of the diffusion coefficient. Much of the theory developed here generalizes results for ordinary diffusions, see, for example, Liptser and Shiryaev (1977), Basawa and Prakasa Rao (1980) and Sørensen (1983), and covers these as special cases. Some particular cases of diffusions with jumps have been studied by Skokan (1984) and Aase and Guttorp (1987). A number of asymptotic statistical problems were studied for a rather general class of diffusions with jumps by Lin'kov (1982).

In Section 2 the mathematics of diffusions with jumps is reviewed, and the type of models considered in the paper is defined. The statistical problem is stated rigorously. An estimator of the diffusion coefficient is proposed. Sufficient conditions are given for the existence of the likelihood function together with an expression in terms of stochastic integrals for the likelihood function. Finally it is shown how the theory simplifies when the jumps are generated by a compound Poisson process. The general likelihood theory is presented in Section 3. It is proved that under mild conditions the score function is a martingale. Various types of observed and expected Fisher information are discussed. The score function is compared to the quasi score function proposed by Hutton and Nelson (1986). The asymptotic distribution of the maximum likelihood estimators and of some test statistics is derived from the martingale properties of the score function. Section 3 is concluded by the result that in important special cases the stochastic integrals in the likelihood function can be expressed in terms of Lebesgue integrals and finite sums. In Section 4 it is demonstrated how the theory simplifies and can be sharpened when a part of the likelihood function has the form of an exponential family. Finally four examples that illustrate various aspects of the theory are discussed in Section 5.

2   DIFFUSIONS WITH JUMPS

Let Ω be the set of all functions $\omega: \mathbb{R}_+ \to \mathbb{R}^d$ that are right continuous with limits from the left. A filtration, that is, an increasing right-continuous sequence of σ-algebras, is defined on Ω by

$$\mathcal{F}_t = \bigcap_{s>t} \sigma(\omega(u) : u \le s)$$

Define $\mathcal{F}$ as the smallest σ-algebra containing $\{\mathcal{F}_t\}$. We shall call $(\Omega, \mathcal{F}, \{\mathcal{F}_t\})$ the canonical space.

Let $(\Omega', \mathcal{F}', \{\mathcal{F}'_t\}, P')$ be a stochastic basis. In particular, $\{\mathcal{F}'_t\}$ is a filtration, and P' is a probability measure on $\mathcal{F}'$. Suppose that this stochastic basis is equipped with an m-dimensional Wiener process W and a class of homogeneous Poisson random measures $\{p^\theta: \theta \in \Theta\}$ on $\mathbb{R}_+ \times \mathbb{R}^k$ (or on $\mathbb{R}_+ \times E$, where E is an arbitrary Blackwell space). The index-set Θ is a subset of $\mathbb{R}^n$ with a nonvoid interior. An integer-valued random measure p on $\mathbb{R}_+ \times \mathbb{R}^k$ is called a Poisson measure if the following two conditions are satisfied. The measure q on $\mathbb{R}_+ \times \mathbb{R}^k$ given by $q(A) = E(p(A))$, $A \in \mathcal{B}(\mathbb{R}_+) \otimes \mathcal{B}(\mathbb{R}^k)$, is σ-finite and satisfies $q(\{t\} \times \mathbb{R}^k) = 0$ for all $t \ge 0$. For all $s \ge 0$ and all $A \in \mathcal{B}(\mathbb{R}_+) \otimes \mathcal{B}(\mathbb{R}^k)$ satisfying $A \subseteq (s, \infty) \times \mathbb{R}^k$, the random variable $p(A)$ is independent of $\mathcal{F}'_s$. The measure q is called the intensity measure of p. Since p^θ is homogeneous, its intensity measure is given by

$$q^\theta(dt, dy) = F_\theta(dy)\, dt \tag{2.1}$$

where F_θ is a σ-finite measure on $(\mathbb{R}^k, \mathcal{B}(\mathbb{R}^k))$. A particularly simple, but important, example is provided by the random measure

$$p(\omega; dt, dy) = \sum_s 1_{\{\Delta Y_s(\omega)\, \neq\, 0\}} \epsilon_{(s, \Delta Y_s(\omega))}(dt, dy) \tag{2.2}$$

Here ϵ_a is the Dirac measure at a, and Y is a k-dimensional compound Poisson process with intensity λ. If the probability distribution of the jumps is G, it follows that

$$q(dt, dy) = \lambda G(dy)\, dt$$

Here q is a finite measure. In general this need not be the case. For instance, an infinite intensity measure is obtained when we let the stochastic process Y in (2.2) be the gamma-process, that is, the process with independent stationary gamma-distributed increments, see Example 5.4.

Now, suppose that, for each $\theta \in \Theta$, the stochastic differential equation

$$dY_t = \beta(\theta; t, Y_t)\, dt + \gamma(t, Y_{t-})\, dWt$$

$$+ \int_{\mathbb{R}^k} \delta(\theta; t, Y_{t-}, z)(p^{\theta}(dt, dz) - q^{\theta}(dt, dz)) \qquad (2.3)$$

$$Y_0 = x_0$$

has a solution Y^{θ}. Here x_0 is a nonrandom d-dimensional vector, while

$$\gamma: \mathbb{R}_+ \times \mathbb{R}^d \to \mathbb{R}^d \times \mathbb{R}^m$$
$$\beta(\theta; \cdot, \cdot): \mathbb{R}_+ \times \mathbb{R}^d \to \mathbb{R}^d$$

and

$$\delta(\theta; \cdot, \cdot, \cdot): \mathbb{R}_+ \times \mathbb{R}^d \times \mathbb{R}^k \to \mathbb{R}^d$$

are Borel functions. Local Lipschitz conditions and linear growth conditions on the coefficients that ensure the existence and uniqueness of a solution of (2.3) can be given, see, for example, Jacod and Shiryaev (1987, Theorem III-2-32). By a solution of (2.3) we mean an $\{\mathcal{F}'_t\}$-adapted d-dimensional stochastic process that is right continuous with limits from the left such that

$$Y_t^{\theta} = x_0 + \int_0^t \beta(\theta; s, Y_s)\, ds + \int_0^t \gamma(s, Y_{s-})\, dW_s$$

$$+ \int_0^t \int_{\mathbb{R}^k} \delta(\theta; s, Y_{s-}, z)(p^{\theta}(ds, dz) - q^{\theta}(ds, dz))$$

It is, of course, assumed that these integrals exist. Note, that if the process corresponding to p^{θ} makes a jump of size z at time t, then Y^{θ} will make a jump of size $\delta(\theta; t, Y_{t-}, z)$ at the same time. The aim of this chapter is to study statistical inference about the parameter θ.

The solution Y^{θ} of (2.3) induces a probability measure P_{θ} on the canonical space $(\Omega, \mathcal{F})$ by the mapping $\omega' \in \Omega' \to Y^{\theta}(\omega') \in \Omega$. Under P_{θ} the canonical process $X_t(\omega) = \omega(t)$, $\omega \in \Omega$, is a semimartingale starting at $X_0 = x_0$ with local characteristics (B, C, ν) given by

$$B_t(\theta) = \int_0^t b(\theta; s, X_s)\, ds \qquad (2.4)$$

$$C_t = \int_0^t c(s, X_s)\, ds \qquad (2.5)$$

and

$$\nu(\theta; dy, dt) = K_t(\theta; X_{t-}, dy)\, dt \qquad (2.6)$$

Here the Borel kernel K_t is defined by

$$K_t(\theta; x, A) = \int_{\mathbb{R}^k} 1_{A \setminus \{0\}}(\delta(\theta; t, x, y)) F_\theta(dy) \qquad (2.7)$$

for all Borel subsets A of $\mathbb{R}^d$, while

$$b(\theta; t, x) = \beta(\theta; t, x) - \int_{\|y\| > 1} y K_t(\theta; x, dy) \qquad (2.8)$$

and

$$c(t, x) = \gamma(t, x)\gamma(t, x)^T \qquad (2.9)$$

We denote transposition of matrices by T. Actually, X is a so-called special semimartingale. This is why the integral in (2.8) always exists.

In the following we will assume that our data are an observation of the canonical process. Obviously, this is not a restriction. We will suppose that a parametric model for the canonical process is given by the local characteristics (2.4)–(2.6) where, for all $\theta \in \Theta$,

$$b(\theta; \cdot, \cdot): \mathbb{R}_+ \times \mathbb{R}^d \to \mathbb{R}^d$$

and

$$c: \mathbb{R}_+ \times \mathbb{R}^d \to \mathbb{R}^d \times \mathbb{R}^d$$

are Borel functions such that the matrix $c(t, x)$ is symmetric and non-negative definite, and $K_t(\theta; x, dy)$ is a Borel transition kernel from $\mathbb{R}_+ \times \mathbb{R}^d$ into $\mathbb{R}^d$ with $K_t(\theta, x, \{0\}) = 0$. A semimartingale with such local characteristics is called a diffusion with jumps. As we have seen, this class of processes covers solutions to equations of the form (2.3). It also covers solutions to the more general type of equations

$$dY_t = b(\theta; t, Y_t) \, dt + \gamma(t, Y_{t-}) \, dW_t$$

$$+ \int_{\|z\| \leq 1} \delta(\theta; t, Y_{t-}, z)(p^\theta(dt, dz) - q^\theta(dt, dz))$$

$$+ \int_{\|z\| > 1} \delta(\theta; t, Y_{t-}, z)p^\theta(dt, dz). \tag{2.10}$$

The solution of such an equation is still a semimartingale, but need not be a special semimartingale. Fewer conditions are needed on δ and q to make (2.10) meaningful than is the case for (2.3). For example, the solution to (2.10) can not be written in the form (2.3) if $q(dt, dz) = f(z) \, dz \, dt$ and $\delta(\theta; t, y, z) = z$, where $f(z) = z^{-(\alpha+1)}$ $(z > 0, 0 < \alpha \leq 1)$ is the density of the Lévy measure for the positive stable process with parameter α. This is because the last integral of (2.3) does not exist. The local characteristics (B, C, ν) are given by (2.4)–(2.7), where b is the coefficient in (2.10).

The assumption that the diffusion coefficient γ does not depend on θ is needed because otherwise the restrictions of P_θ and P_{θ_1} $(\theta \neq \theta_1)$ to $\mathcal{F}_t$ would have been singular for all $t > 0$. However, as is the case for ordinary diffusions, we do have a means for estimating γ when it is not known. Under P_θ the matrix

$$\sum_{j=1}^{2^n} (X_{js2^{-n}} - X_{(j-1)s2^{-n}})(X_{js2^{-n}} - X_{(j-1)s2^{-n}})^T$$

$$- \sum_{u \leq s} (\Delta X_u)(\Delta X_u)^T$$

converges to C_s in probability as $n \to \infty$ uniformly for $s \leq t$. Therefore, we can estimate C_s by this matrix for n large enough. Note that, as one would expect, the fluctuation of the diffusion is overestimated if a model without jumps is used in a situation where jumps do occur. This can, in some applications, be crucial, see Example 5.3.

An account of the theory of diffusions with jumps can be found in Jacod (1979, Chapter 13). See also Jacod and Shiryaev (1987). Diffusions with jumps are Markov processes. The generator of such a process can be found in Stroock (1975) or Jacod (1979).

We shall need the following condition on triples (b, c, K) of two Borel functions and a Borel transition kernel of the kind defined above.

Condition C

(a) b is bounded on $[0, n] \times \{x: \|x\| \le n\}$ for all $n \in \mathbb{N}$.
(b) c is continuous on $\mathbb{R}_+ \times \mathbb{R}^d$ and everywhere strictly positive definite.
(c) The function

$$(t, x) \to \int_A (|y|^2 \wedge 1) K_t(x, dy)$$

is continuous on $\mathbb{R}_+ \times \mathbb{R}^d$ for all Borel subsets A of $\mathbb{R}^d$.

In the rest of the chapter we shall assume that, for each $\theta \in \Theta$, the triple $(b(\theta), c, K(\theta))$ satisfies Condition C. This assumption implies that, for each $\theta \in \Theta$, there exists a unique probability measure P_θ on $(\Omega, \mathcal{F})$ under which the canonical process X is a diffusion with jumps with local characteristics $(B(\theta), C, \nu(\theta))$ given by (2.4)–(2.6). Under P_θ the process

$$X_t^c(\theta) = X_t - x_0 - \sum_{s \le t} \Delta X_s 1_{\{\|\Delta X_s\| > 1\}}$$

$$- B_t(\theta) - \int_0^t \int_{\|x\| \le 1} x(\mu - \nu(\theta))(dx, ds) \qquad (2.11)$$

is a continuous local martingale. The random measure μ is defined by

$$\mu(\omega; dx, dt) = \sum_s 1_{\{\Delta X_s(\omega) \ne 0\}} \epsilon_{(s, \Delta X_s(\omega))}(dt, dx) \qquad (2.12)$$

where ϵ_a denotes the Dirac measure at a. For a definition of a stochastic integral with respect to $\mu - \nu(\theta)$, see Jacod (1979). We will also assume that for some fixed $\theta^{(0)} \in \Theta$

$$K_t(\theta; x, dy) = Y(\theta; t, x, y) K_t(\theta^{(0)}; x, dy) \qquad (2.13)$$

where $(t, x, y) \to Y(\theta; t, x, y)$ is a strictly positive function from $\mathbb{R}_+ \times \mathbb{R}^d \times \mathbb{R}^d$ into $\mathbb{R}_+$.

We shall be concerned with data obtained by observing the process X permanently in the time interval $[0, t]$. Therefore, the relevant probability measures are the restrictions P_θ^t of P_θ to the σ-algebra $\mathcal{F}_t$. We shall now give a theorem with sufficient conditions that $P_\theta^t \sim P_{\theta^{(0)}}^t$ and with an expression for the likelihood function $L_t(\theta) = dP_\theta^t/dP_{\theta^{(0)}}^t$.

Define, for each $\theta \in \Theta$, a d-dimensional process $y(\theta)$ by

$$y_t(\theta) = b(\theta; t, X_{t-}) - b(\theta^{(0)}; t, X_{t-})$$

$$- \int_{\|y\| \le 1} y[Y(\theta; t, X_{t-}, y) - 1]K_t(\theta^{(0)}; X_{t-}, dy) \quad (2.14)$$

and another process by

$$\alpha_t(\theta) = c_t^{-1} y_t(\theta)$$

where $c_t = c(t, X_{t-})$. In case the integral in (2.14) diverges, we set y_t and α_t equal to $+\infty$. We also need the increasing nonnegative real process

$$A_t(\theta) = \int_0^t \alpha_s(\theta)^T c_s \alpha_s(\theta) \, ds$$

$$+ \int_0^t \int_{\mathbb{R}^d} [Y(\theta; s, X_s, y) - 1]1_{\{Y > 2\}} K_s(\theta^{(0)}; X_s, dy) \, ds$$

$$+ \int_0^t \int_{\mathbb{R}^d} [Y(\theta; s, X_s, y) - 1]^2 1_{\{Y \le 2\}} K_s(\theta^{(0)}; X_s, dy) \, ds$$

$$(2.15)$$

which may take the value ∞. This process is continuous except possibly at a point where it jumps to infinity.

Theorem 2.1. Suppose

$$P_\theta(A_t(\theta) < \infty) = P_{\theta^{(0)}}(A_t(\theta) < \infty) = 1$$

Then

$$P_\theta^t \sim P_{\theta^{(0)}}^t$$

and

$$\frac{dP_\theta^t}{dP_{\theta^{(0)}}^t} = \exp\left\{ \int_0^t y_s(\theta)^T c_s^{-1} \, dX_s^c(\theta^{(0)}) \right.$$

$$- \frac{1}{2} \int_0^t y_s(\theta)^T c_s^{-1} y_s(\theta) \, ds$$

$$+ \int_0^t \int_{\mathbb{R}^d} [Y(\theta; s, X_{s-}, y) - 1](\mu - \nu(\theta^{(0)}))(dy, ds)$$

$$+ \left. \int_0^t \int_{\mathbb{R}^d} [\log(Y(\theta; s, X_{s-}, y)) - Y(\theta; s, X_{s-}, y) + 1]\mu(dy, ds) \right\}$$

$$(2.16)$$

where $X^c(\theta^{(0)})$ is the continuous $P_{\theta^{(0)}}$-martingale given by (2.11), and μ is the random measure defined by (2.12).

Proof. The result follows from Theorem 4.2 and Theorem 4.5(b) in Jacod and Mémin (1976). To see this, note that

$$B_t(\theta) = B_t(\theta^{(0)}) + \int_0^t c_s \alpha_s(\theta) \, ds$$

$$+ \int_0^t \int_{\|y\| \leq 1} y[Y(\theta; s, X_{s-}, y) - 1] \nu(\theta^{(0)}; dy, ds)$$

Define a stopping time by

$$\tau = t \wedge \inf\{s: A_s(\theta) \geq n\}$$

and the stopped process X^τ by $X_s^\tau = X_{s \wedge \tau}$. Under Condition C any two probability measures P_θ and P_θ' under both of which X^τ is a semimartingale with local characteristics $(B^\tau(\theta), C^\tau, \nu^\tau(\theta))$ coincide on $\mathscr{F}_\tau^0$. Here $\mathscr{F}_s^0 = \sigma(X_u: \ u \leq s)$ and

$$\nu^\tau(\theta; dy, ds) = \nu(\theta; dy, ds) 1_{\{s \leq \tau\}}$$

Hence the conditions in the above-mentioned theorems are satisfied. The existence under P of the integrals in the expression for the Radon–Nikodym derivative follows from the finiteness of $A_t(\theta)$.

Instead of $A_t(\theta)$ one can in Theorem 2.1 equally well use the process

$$\tilde{A}_t(\theta) = \int_0^t \alpha_s(\theta)^T c_s \alpha_s(\theta) \, ds$$

$$+ \int_0^t \int_{\mathbb{R}^d} (1 - Y(\theta; s, X_s, y)^{1/2})^2 K_s(\theta^{(0)}; X_s, dy) \, ds$$

The theorems in Jacod and Mémin (1976) are only formulated for one-dimensional processes, but can be generalized. Theorem 2.1 also follows from Theorem III-5-34 in Jacod and Shiryaev (1987). This theorem explicitly covers d-dimensional processes.

We conclude this section by considering an important special case. Suppose that the σ-finite measure F_θ in (2.1) has a density $f(\theta; y)$ with respect to the Lebesgue measure on $\mathbb{R}^k$ and that all F_θ, $\theta \in \Theta$, are equivalent. Assume, moreover, that $k = d$ and that, for each (θ, t, x), the function

$$z \to \delta(\theta; t, x, z)$$

has a continuously differentiable inverse

$$y \to \varphi(\theta; t, x, y)$$

with Jacobian $|J_\varphi(\theta; t, x, y)| \neq 0$ for all $y \in \text{Im}(\delta) = \delta(\theta; t, x, \mathbb{R}^d)$. Under these assumptions

$$K_t(\theta; x, A) = \int_{A \cap \text{Im}(\delta)} f(\theta; \varphi(\theta; t, x, y)|J_\varphi(\theta; t, x, y)| \, dy$$

so that (2.13) is satisfied with

$$Y(\theta; t, x, y) = \frac{f(\theta; \varphi(\theta; t, x, y))|J_\varphi(\theta; t, x, y)|}{f(\theta^{(0)}; \varphi(\theta^{(0)}; t, x, y))|J_\varphi(\theta^{(0)}; t, x, y)|}$$

Let us specialize further and assume that the Poisson measure p^θ is generated by a compound Poisson process with intensity $\lambda(\theta)$, see (2.2). If the distribution of the jumps has density $g(\theta; y)$ with respect to the Lebesgue measure on $\mathbb{R}^d$, then $f(\theta; y) = \lambda(\theta)g(\theta; y)$, so F_θ is a bounded measure. Consequently, the jump mechanism does not cause problems when proving the existence of the likelihood function. This is reflected in the fact that the process $A(\theta)$ defined by (2.15) is dominated in the following way:

$$A_t(\theta) \leq t(\lambda(\theta) + 2\lambda(\theta_0)) + \int_0^t y_s(\theta)^T c_s^{-1} y_s(\theta) \, ds \qquad (2.17)$$

As a consequence the conditions in Theorem 2.1 simplify to conditions similar to those for the existence of the likelihood function in the case of ordinary diffusions. Also, the expression for the Radon–Nikodym derivative simplifies considerably. We formulate this result in the following corollary.

Corollary 2.2. Let F_θ in (2.1) be given by $F_\theta(dy) = \lambda(\theta)g(\theta; y) \, dy$ for some equivalent probability density functions $g(\theta; y)$. Let $d = k$ and assume the conditions above on δ. Then

$$P_\theta \left(\int_0^t \alpha_s(\theta)^T c_s \alpha_s(\theta) \, ds < \infty \right) = P_{\theta^{(0)}} \left(\int_0^t \alpha_s(\theta)^T c_s \alpha_s(\theta) \, ds < \infty \right) = 1$$

implies that

$$P_\theta^t \sim P_{\theta^{(0)}}^t$$

and that

$$\frac{dP^t_\theta}{dP^t_{\theta^{(0)}}} = \exp\left\{ \int_0^t y_s(\theta)^T c_s^{-1}\, dX_s^c(\theta^{(0)}) - \frac{1}{2}\int_0^t y_s(\theta)^T c_s^{-1} y_s(\theta)\, ds \right.$$

$$\left. + \sum_{s\in S_t} \log(Y(\theta; s, X_{s-}, \Delta X_s)) + (\lambda(\theta^{(0)}) - \lambda(\theta))t \right\}$$

where $S_t = \{s \le t : \Delta X_s \ne 0\}$.

Omitting terms independent of θ, we see that the log-likelihood function can be written in the form

$$l_t(\theta) = \int_0^t y_s(\theta)^T c_s^{-1}\, dX_s^c(\theta^{(0)}) - \frac{1}{2}\int_0^t y_s(\theta)^T c_s^{-1} y_s(\theta)\, ds$$

$$+ N_t \log[\lambda(\theta)] - t\lambda(\theta)$$

$$+ \sum_{s\in S_t} \log[g(\theta; \varphi(\theta; s, X_{s-}, \Delta X_s))|J_\varphi(\theta; s, X_{s-}, \Delta X_s)|] \qquad (2.18)$$

where N_t is the Poisson process that counts the number of jumps before time t. Note that the log-likelihood function splits nicely into three components: a diffusion part, a Poisson process part, and a jump-size part.

When the jumps are generated by a compound Poisson process Z^θ, it is natural to write the stochastic differential equation (2.3) in the form

$$dY_t = d(\theta; t, Y_t)\, dt + \gamma(t, Y_{t-})\, dW_t + \tilde\delta(\theta; t, Y_{t-}, \Delta Z_t^\theta)\, dZ_t^\theta \quad (2.19)$$

This is always possible because there is only a finite number of jumps in finite time intervals. With this notation the process $y_t(\theta)$ is given by

$$y_t(\theta) = d(\theta; t, X_{t-}) - d(\theta^{(0)}; t, X_{t-})$$

rather than by (2.14), and $\delta(\theta; t, x, z) = \tilde\delta(\theta; t, x, z)z$. Fewer conditions are needed on δ and Z to make (2.19) meaningful than is the case for (2.3). With the notation of (2.19), the continuous P_θ-martingale $X^c(\theta)$ is given by

$$X_t^c(\theta) = X_t - x_0 - \sum_{s\le t} \Delta X_s - \int_0^t d(\theta; s, X_s)\, ds \qquad (2.20)$$

Example 2.1. Consider the compound Poisson process case, and assume that $k = d = 1$ and that

$$\delta(\theta; t, x, z) = z$$

Then the jumps of our processes equal those of the compound Poisson process. In this case

$$Y(\theta; t, x, y) = \frac{\lambda(\theta)g(\theta; y)}{\lambda(\theta^{(0)})g(\theta^{(0)}; y)}$$

and $K(\theta)$ satisfies Condition C trivially.

Example 2.2. Still in the compound Poisson process case with $k = d = 1$, suppose that

$$\delta(\theta; t, x, z) = z\rho(\theta; t, x)$$

where

$$\rho(\theta; t, x) = 1 - \exp(-\psi(\theta, t)|x|)$$

for some continuous function $\psi > 0$ of θ and t. Here the jumps die away when the process is close to zero, but at a distance from zero depending on θ and time the process makes jumps almost like the compound Poisson process. When $x \neq 0$

$$\varphi(\theta; t, x, y) = y/\rho(\theta; t, x)$$

and

$$|J_\varphi(\theta; t, x, y)| = \rho(\theta; t, x)^{-1}$$

Hence

$$Y(\theta; t, x, y) = \begin{cases} \dfrac{\lambda(\theta)g(\theta; y/\rho(\theta; t, x))\rho(\theta^{(0)}; t, x)}{\lambda(\theta^{(0)})g(\theta^{(0)}; y/\rho(\theta^{(0)}; t, x))\rho(\theta; t, x)} & \text{for } x \neq 0 \\ 1 & \text{for } x = 0 \end{cases}$$

If $g(\theta; y)$ decreases monotonically as $|y| \to \infty$ for $|y|$ large enough, and if the νth moment of $g(\theta; y)$ exists for some $\nu > 0$, it is easy to show that $K(\theta)$ satisfies Condition C. The inequality (2.17) is also satisfied in this case, where $z \to \delta(\theta; t, x, z)$ is not one-to-one when $x = 0$. Hence the result of Corollary 2.2 holds.

3 LIKELIHOOD THEORY

Under the conditions of Theorems 2.1 the log-likelihood function is given by

$$l_t(\theta) = \int_0^t y_s(\theta)^T c_s^{-1}\, dX_s^c(\theta^{(0)}) - \frac{1}{2}\int_0^t y_s(\theta)^T c_s^{-1} y_s(\theta)\, ds$$

$$+ \int_0^t \int_{\mathbb{R}^d} [Y(\theta; s, X_{s-}, y) - 1](\mu - \nu(\theta^{(0)}))(dy, ds)$$

$$+ \int_0^t \int_{\mathbb{R}^d} [\log(Y(\theta; s, X_{s-}, y)) - Y(\theta; s, X_{s-}, y) + 1]\mu(dy, ds)$$

$$(3.1)$$

Provided $l_t(\theta)$ is differentiable with respect to θ and that differentiation with respect to θ and stochastic integration can be interchanged, we find the following expression for the score vector

$$\dot{l}_t(\theta) = \int_0^t \dot{y}_s(\theta)^T c_s^{-1}\, dX_s^c(\theta^{(0)}) - \int_0^t \dot{y}_s(\theta)^T c_s^{-1} y_s(\theta)\, ds$$

$$+ \int_0^t \int_{\mathbb{R}^d} \dot{Y}(\theta; s, X_{s-}, y)(\mu - \nu(\theta^{(0)}))(dy, ds)$$

$$+ \int_0^t \int_{\mathbb{R}^d} [\dot{H}(\theta; s, X_{s-}, y) - \dot{Y}(\theta; s, X_{s-}, y)]\mu(dy, ds) \qquad (3.2)$$

In particular, it has been assumed that these integrals exist. The matrix $\dot{y}_s(\theta)$ is given by $\dot{y}_s(\theta)_{ij} = \partial y_s(\theta)_i/\partial \theta_j$, the vector $\dot{Y}(\theta; s, x, y)$ by $\dot{Y}(\theta; s, x, y)_i = \partial Y(\theta; s, x, y)/\partial \theta_i$, and $H(\theta; s, x, y) = \log(Y(\theta; s, x, y))$. It should, of course, be checked in particular models that the interchange of differentiation and integration is allowed. Results about interchanging differentiation and stochastic integration are given in Karandikar (1983) and Hutton and Nelson (1984a). In the important case that p^θ corresponds to a compound Poisson process, the last two integrals in (3.1) simplify to a finite sum, see Corollary 2.2, which causes no problems. In this situation also the integral with respect to $X^c(\theta^{(0)})$ is usually painless provided X is one-dimensional (see the discussion at the end of this section).

Now, suppose

$$P_\theta\left(\int_0^t \int_{\mathbb{R}^d} |\dot{Y}(\theta; s, X_s, y)_i| \nu(\theta^{(0)}; dy, ds) < \infty\right) = 1$$

for all $i = 1, \ldots, n$, all $\theta \in \Theta$ and all $t \geq 0$. Then

$$l_t(\theta) = \int_0^t \dot{y}_s(\theta)^T c_s^{-1} \, dX_s^c(\theta)$$

$$+ \int_0^t \int_{\mathbb{R}^d} \dot{H}(\theta; s, X_{s-}, y)(\mu - \nu(\theta))(dy, ds) \qquad (3.3)$$

where, under P_θ,

$$X_t^c(\theta) = X_t^c(\theta^{(0)}) - \int_0^t y_s(\theta) \, ds$$

is a continuous local martingale with quadratic characteristic C given by (2.5). The measure $\nu(\theta)$, defined by (2.6), is the dual predictable compensator of μ under P_θ. Hence the score vector is under P_θ a locally square integrable martingale provided the quadratic characteristic

$$\langle l(\theta) \rangle_t = \int_0^t \dot{y}_s(\theta)^T c_s^{-1} \dot{y}_s(\theta) \, ds$$

$$+ \int_0^t \int_{\mathbb{R}^d} \dot{H}(\theta; s, X_s, y) \dot{H}(\theta; s, X_s, y)^T \nu(\theta; dy, ds) \qquad (3.4)$$

is almost surely finite for all t. In the calculation of (3.4) we have used that the first integral in (3.3) is a continuous martingale, while the second is a purely discontinuous martingale. If, moreover, $E_\theta(\langle l(\theta) \rangle_t) < \infty$ for all $t \geq 0$, the score function is a square integrable martingale under P_θ.

The quadratic variation of $l_t(\theta)$ is

$$[l(\theta)]_t = \int_0^t \dot{y}_s(\theta)^T c_s^{-1} \dot{y}_s(\theta) \, ds$$

$$+ \int_0^t \int_{\mathbb{R}^d} \dot{H}(\theta; s, X_{s-}, y) \dot{H}(\theta; s, X_{s-}, y)^T \mu(dy, ds) \qquad (3.5)$$

The matrix $I_t(\theta) = \langle l(\theta) \rangle_t$ is the incremental expected information, while $J_t(\theta) = [l(\theta)]_t$ is the incremental observed information matrix. The difference $J_t(\theta) - I_t(\theta)$ is a matrix of local P_θ-martingales. If $E_\theta(I_t(\theta)) < \infty$ for all $t \geq 0$, the difference is a matrix of zero-mean martingales. For a discussion of the various types of Fisher information for stochastic processes, see Barndorff–Nielsen and Sørensen (1989).

In order to calculate the observed Fisher information, $j_t(\theta)$, we suppose that the log-likelihood function is twice differentiable, and that we can interchange differentiation and integration again. We find that

$$j_t(\theta) = -\ddot{l}_t(\theta)$$

$$= J_t(\theta) - \int_0^t \ddot{y}_s(\theta)^T c_s^{-1} \, dX_s^c(\theta)$$

$$- \int_0^t \int_{\mathbb{R}^d} \frac{\ddot{Y}(\theta; s, X_{s-}, y)}{Y(\theta; s, X_{s-}, y)} (\mu - \nu(\theta))(dy, ds)$$

$$= I_t(\theta) - \int_0^t \ddot{y}_s(\theta)^T c_s^{-1} \, dX_s^c(\theta)$$

$$- \int_0^t \int_{\mathbb{R}^d} \ddot{H}(\theta; s, X_{s-}, y)(\mu - \nu(\theta))(dy, ds) \qquad (3.6)$$

provided

$$P_\theta \left(\int_0^t \int_{\mathbb{R}^d} |\ddot{Y}(\theta; s, X_s, y)_{ij}| \nu(\theta^{(0)}; dy, ds) < \infty \right) = 1$$

for all $i, j = 1, \ldots, n$, all $\theta \in \Theta$ and all $t \geq 0$. The $d \times n^2$ matrix $\ddot{y}_s(\theta)$ is given by $\ddot{y}_s(\theta)_{i,(k,l)} = \partial^2 y_s(\theta)_i / \partial \theta_k \partial \theta_l$, and the $n \times n$ matrix $\ddot{Y}$ by $\ddot{Y}(\theta; t, x, y)_{ij} = \partial^2 Y(\theta; t, x, y) / \partial \theta_i \partial \theta_j$. If, moreover,

$$E_\theta \left(\int_0^t \ddot{y}_s(\theta)^T c_s^{-1} \ddot{y}_s(\theta) \, ds \right) < \infty$$

and

$$E_\theta \left(\int_0^t \int_{\mathbb{R}^d} \ddot{Y}(\theta; s, X_s, y) \nu(\theta^{(0)}; dy, ds) \right) < \infty$$

for all $t > 0$, then the difference $j_t(\theta) - J_t(\theta)$ is a matrix of zero-mean martingales under P_θ. In particular,

$$i_t(\theta) = E_\theta(j_t(\theta)) = E_\theta(I_t(\theta)) = E_\theta(J_t(\theta))$$

provided $E_\theta(I_t(\theta)) < \infty$. Here $i_t(\theta) = E_\theta(\dot{l}_t(\theta)\dot{l}_t(\theta)^T)$ is the expected Fisher information. Note that the results about the score vector and about Fisher information are completely analogous to results for ordinary diffusions and diffusion-type processes (see Sørensen (1983)).

It is of some interest to compare the score vector (3.3) to the quasi-score vector proposed by Hutton and Nelson (1986) [see also Godambe and Heyde (1987)]. Suppose X is a solution to an equation of the type (2.3) and thus is a special semimartingale. Then the conditions of Hutton and Nelson are satisfied, and their quasi score vector is given by

$$\int_0^t \dot{\beta}(\theta; s, X_s)^T \left[c_s + \int_{\mathbb{R}^d} yy^T K_s(\theta; X_s, dy) \right]^{-1} dM_s(\theta) \qquad (3.7)$$

where the local martingale $M(\theta)$ is given by

$$M_t(\theta) = X_t^c(\theta) + \int_0^t \int_{\mathbb{R}^d} y(\mu - \nu(\theta))(dy, ds)$$

Note that

$$\dot{\beta}(\theta; t, X_t) = \dot{y}_t(\theta) + \int_{\mathbb{R}^d} z \dot{Y}(\theta; t, X_t, z)^T K_t(\theta^{(0)}; X_t, dz)$$

so, while the score function (3.3) is a sum of two terms, one connected to the diffusion and the other to the jumps, these two terms are thoroughly mixed in the quasi score function. The difference between (3.3) and (3.7) suggest an alternative to the Hutton-Nelson quasi score vector. This point has been pursued in Sørensen (1990).

The next theorem is concerned with the asymptotic properties of the score vector. For a positive semidefinite matrix A, denote by $A^{1/2}$ the unique positive semidefinite square root of A, and by $\det(A)$ the determinant of A. Let I_n be the $n \times n$ identity marix and $\text{diag}(x_1, \ldots, x_n)$ the $n \times n$ diagonal matrix with $x_1, \ldots, x_n$ in the diagonal.

Theorem 3.1. Suppose all diagonal elements of $i_t(\theta)$ tend to infinity as $t \to \infty$ and that, as $t \to \infty$,

$$i_t(\theta)_{jj}^{-1/2} E_\theta(\sup_{s \le t} |\dot{H}(\theta; s, X_{s-}, \Delta X_s)_j| l_{\{\Delta X_s \ne 0\}}) \to 0 \qquad j = 1, \ldots, n$$

$$(3.8)$$

$$D_t(\theta)^{-1/2} J_t(\theta) D_t(\theta)^{-1/2} \to \eta^2(\theta) \text{ in probability under } P_\theta \qquad (3.9)$$

and

$$D_t(\theta)^{-1/2} i_t(\theta) D_t(\theta)^{-1/2} \to \Sigma(\theta) \qquad (3.10)$$

where $D_t(\theta) = \text{diag}(i_t(\theta)_{11}, \ldots, i_t(\theta)_{nn})$, $\eta^2(\theta)$ is a random positive semidefinite matrix, and $\Sigma(\theta)$ is a nonrandom positive definite matrix.

Then we have the following results about convergence in distribution under P_θ as $t \to \infty$:

$$D_t(\theta)^{-1/2} \dot{l}_t(\theta) \to Z \quad \text{(stably)} \tag{3.11}$$

and conditionally on $\det(\eta^2(\theta)) > 0$

$$[D_t(\theta)^{1/2} J_t(\theta)^{-1} D_t(\theta)^{1/2}]^{1/2} D_t(\theta)^{-1/2} \dot{l}_t(\theta) \to N(0, I_n) \quad \text{(mixing)} \tag{3.12}$$

and

$$\dot{l}_t(\theta)^T J_t(\theta)^{-1} \dot{l}_t(\theta) \to \chi^2(n) \quad \text{(mixing)} \tag{3.13}$$

Here the distribution of Z is the normal variance mixture with characteristic function $\varphi(u) = E_\theta(\exp(-\frac{1}{2} u^T \eta^2(\theta) u))$, $u = (u_1, \ldots, u_n)^T$.

Proof. Since

$$\Delta \dot{l}_t(\theta) = \dot{H}(\theta; t, X_{t-}, \Delta X_t) 1_{\{\Delta X_t \neq 0\}}$$

the conditions of the central limit theorem for square integrable martingales given in the Appendix are satisfied for $\dot{l}_t(\theta)$.

The result (3.13) gives the asymptotic distribution of a score-test statistic for the point hypothesis $\theta = \bar{\theta}$ when the hypothesis is true and can be used to construct confidence regions for θ. Note, that (3.10) is trivially satisfied when $i_t(\theta)$ is a diagonal matrix. When the diagonal elements of $i_t(\theta)$ tend to infinity at the same rate, it follows from Theorem A.1 that

$$J_t(\theta)^{-1/2} \dot{l}_t(\theta) \to N(0, I_n) \quad \text{(mixing)} \tag{3.14}$$

conditionally on $\det(\eta^2(\theta)) > 0$.

Jacod and Shiryaev (1987, Chapter 8) give a central limit theorem for local martingales without the assumption of square integrability. This suggests the possibility of obtaining a result like Theorem 3.2 under weaker assumptions than those imposed here. A central limit theorem with convergence results that are uniform in θ, but which only cover the nonergodic case (that is, $\eta^2(\theta)$ nonrandom), is given in Lin'kov (1983).

In the very general setup considered here general results about existence and uniqueness of a consistent maximum likelihood estimator for θ will usually not be very helpful in concrete models. See, however, Sørensen (1989a) for a result of this type. The more general results in

Sørensen (1989b) apply to maximum likelihood estimators for diffusions with jumps as a special case. See also Barndorff-Nielsen and Sørensen (1989). Here we will suppose that for t large enough a solution $\hat{\theta}_t$ exists of the likelihood equation $l_t(\theta) = 0$, and that $\hat{\theta}_t \to \theta$ in probability under P_θ as $t \to \infty$. Then we have the following result.

Corollary 3.2. Suppose the conditions of Theorem 3.1 are satisfied and that

$$D_t(\theta)^{-1/2} j_t(\tilde{\theta}_t) D_t(\theta)^{-1/2} \to \eta^2(\theta) \tag{3.15}$$

in probability under P_θ as $t \to \infty$ for any sequence of n-dimensional random vectors $\tilde{\theta}_t$ lying on the line segment connecting θ and $\hat{\theta}_t$. Then conditionally on $\det(\eta^2(\theta)) > 0$

$$D_t(\theta)^{1/2}(\hat{\theta}_t - \theta) \to \eta(\theta)^{-2} Z \quad \text{(stably)} \tag{3.16}$$

$$[D_t(\theta)^{-1/2} j_t(\hat{\theta}_t) D_t(\theta)^{-1/2}]^{1/2} D_t(\theta)^{1/2}(\hat{\theta}_t - \theta) \to N(0, I_n) \quad \text{(mixing)} \tag{3.17}$$

and

$$2(l_t(\hat{\theta}_t) - l_t(\theta)) \to \chi^2(n) \quad \text{(mixing)} \tag{3.18}$$

in distribution under P_θ as $t \to \infty$.

Proof. By series expansion of $l_t(\theta)$ we see that

$$l_t(\theta) = j_t(\tilde{\theta}_t)(\hat{\theta}_t - \theta)$$

for some $\tilde{\theta}_t$ on the line segment connecting θ and $\hat{\theta}_t$. Similarly,

$$l_t(\hat{\theta}_t) - l_t(\theta) = \tfrac{1}{2}(\hat{\theta}_t - \theta)^T j_t(\tilde{\theta}_t)(\hat{\theta}_t - \theta)$$

From these expansions the theorem follows immediately. To obtain (3.16) the stability of (3.11) is used.

Note that the measure of the dispersion of $\hat{\theta}_t$ around θ obtained from (3.17) is simply $j_t(\hat{\theta}_t)^{-1}$. Therefore, it does not matter in applications that $D_t(\theta)$ depends on θ.

Corollary 3.3. Suppose the conditions of Theorem 3.1 and (3.15) are satisfied, and that the components of $D_t(\theta)$ tend to infinity at the same rate. Then conditionally on $\det(\eta^2(\theta)) > 0$

$$j_t(\hat{\theta}_t)^{1/2}(\hat{\theta}_t - \theta) \to N(0, I_n) \quad \text{(mixing)} \tag{3.19}$$

in distribution under P_θ as $t \to \infty$.

Proof. Use Theorem A.1 with $k_{1t} = \cdots = k_{nt} = i_t(\theta)_{11}^{1/2}$.

We shall complete this section by showing how the first stochastic integral in the log-likelihood function (3.1) can be expressed in terms of Lebesgue integrals and a finite sum when the diffusion with jumps is one-dimensional and the jumps are generated by a compound Poisson process. This generalizes well-known results for ordinary diffusions. Define a nonrandom function by

$$F(x, t) = \int_{x_0}^{x} y_t(\theta; z) c_t(z)^{-1} \, dz$$

where we indicate the dependence of y_t and c_t on X_t by $y_t(\theta; X_{t-})$ and $c_t(X_{t-})$. Suppose that $y_t(\theta; z)$ and $c_t(z)$ are continuously differentiable functions of t and z. Then, by Ito's formula for semimartingales,

$$\int_0^t y_s(\theta; X_{s-}) c_s(X_{s-})^{-1} \, dX_s$$

$$= F(X_t, t) - \sum_{s \le t} [F(X_s, s) - F(X_{s-}, s) - y_s(\theta; X_{s-}) c_s(X_{s-})^{-1} \Delta X_s]$$

$$- \int_0^t \left[\frac{1}{2} \frac{\partial y_s}{\partial z}(\theta; X_s) - \tfrac{1}{2} y_s(\theta; X_s) \frac{\partial}{\partial z}(\ln c_s(X_s)) - \frac{\partial F}{\partial s}(X_s, s) \right] ds$$

On the other hand, $X_t^c(\theta^{(0)})$ is given by (2.20). Hence

$$\int_0^t y_s(\theta; X_{s-}) c_s(X_{s-})^{-1} \, dX_s^c(\theta^{(0)})$$

$$= F(\theta; X_t, t) - \sum_{s \le t} [F(\theta; X_s, s) - F(\theta; X_{s-}, s)]$$

$$- \int_0^t \left[\frac{1}{2} \frac{\partial y_s}{\partial z}(\theta; X_s) - \tfrac{1}{2} y_s(\theta; X_s) \frac{\partial}{\partial z}(\ln c_s(X_s)) - \frac{\partial F}{\partial s}(\theta; X_s, s) \right] ds$$

$$- \int_0^t y_s(\theta; X_s) c_s(X_s)^{-1} \, d(\theta^{(0)}; s, X_s) \, ds \tag{3.20}$$

where we have now emphasized the dependence of F on θ.

The expression (3.20) for the stochastic integral is important for two reasons. First the Riemann sum approximations to the stochastic integral that can be constructed from a sample path will, in general, only converge in probability to the integral (Jacod and Shiryaev, 1987, Proposition I-4-44). Therefore the above explicit expression is preferable.

Secondly, the expression (3.20) gives a very useful result about the interchange in (3.1) of differentiation with respect to θ and stochastic integration. To differentiate the right-hand side of (3.20) we just need results for Lebesgue integrals. If, for instance, $\partial y/\partial\theta$, $\partial^2 y/\partial\theta\,\partial z$, and $\partial^2 y/\partial\theta\,\partial t$ exist and are continuous functions of (θ, t, z), then we can differentiate under the Lebesgue integrals. To see that the score function is given by (3.2) we just give the above arguments with y replaced by $\partial y/\partial\theta$.

A result analogous to (3.20) can not in general be found for multi-dimensional processes. However, in case the ith component of $y_s(\theta; X_s)^T c_s(X_s)^{-1}$ depends on the ith component of X_s only, the preceding arguments generalize immediately.

A result similar to (3.20) can be proved without assumptions on the generation of the jumps. The result, which is not quite as useful as (3.20), is that

$$\int_0^t y_s(\theta; X_{s-})c_s(X_{s-})^{-1}\,dX_s^c(\theta^{(0)})$$

$$= F(\theta; X_t, t) - \int_0^t y_s(\theta; X_s)c_s(X_s)^{-1}b(\theta^{(0)}; s, X_s)\,ds$$

$$- \int_0^t \left\{ \frac{1}{2}\frac{\partial y_s}{\partial z}(\theta; X_s) - \tfrac{1}{2}y_s(\theta; X_s)\frac{\partial}{\partial z}\ln[c_s(X_s)] - \frac{\partial F}{\partial s}(\theta; X_s, s) \right\}ds$$

$$- \sum_{s \le t} [F(\theta; X_s, s) - F(\theta; X_{s-}, s)]1_{\{|\Delta X_s| > 1\}}$$

$$- \sum_{s \le t} [F(\theta; X_s, s) - F(\theta; X_{s-}, s) - y_s(\theta; X_{s-})c_s(X_{s-})^{-1}\Delta X_s]1_{\{|\Delta X_s| \le 1\}}$$

$$- \int_0^t \int_{|x| \le 1} xy_s(\theta; X_{s-})c_s(X_{s-})^{-1}(\mu - \nu(\theta^{(0)}))(dx, ds).$$

Here b is defined by (2.8).

4 EXPONENTIAL FAMILIES

In this section we shall consider how the theory simplifies and can be sharpened in some important cases where the likelihood function, or a part of it, has the form of an exponential family. This generalizes similar results for ordinary diffusions; see, for example, Sørensen (1983). A comprehensive discussion of when a statistical semimartingale model is an exponential family can be found in Küchler and Sørensen (1989).

We will assume that the parameter consists of two parts θ and κ of dimension n_1 and n_2, respectively, which vary independently. Moreover, we will suppose that the process y_t defined by (2.14) depends on θ only, while Y introduced in (2.13) depends on κ only. The process y_t is essentially the part of the drift that is not due to the jumps. Under these assumptions the likelihood function splits into two factors depending on θ and κ, respectively. Hence θ and κ are observed orthogonal. All the information matrices mentioned in Section 3 split into two blocks of size $n_1 \times n_1$ and $n_2 \times n_2$, respectively, in the diagonal with zeros outside. The two blocks depend on θ and κ, respectively.

We shall consider models for which

$$y_t(\theta) = a_t + D_t \theta \tag{4.1}$$

where a_t is a d-dimensional stochastic process, while D_t is a $d \times n_1$ matrix of stochastic processes. Suppose the likelihood function exists. Then the part of the log-likelihood function that depends on θ is given by

$$l_t^{(1)}(\theta) = \theta^T H_t - \tfrac{1}{2}\theta^T \int_0^t D_s^T c_s^{-1} D_s \, ds\, \theta \tag{4.2}$$

where

$$H_t = \int_0^t D_s^T c_s^{-1} \, dX_s^c(\theta^{(0)}, \kappa^{(0)}) - \int_0^t D_s^T c_s^{-1} a_s \, ds \tag{4.3}$$

Here $(\theta^{(0)}, \kappa^{(0)})$ is the parameter value corresponding to the dominating probability measure. The score vector derived from (4.2) is

$$\dot{l}_t^{(1)}(\theta) = H_t - \int_0^t D_s^T c_s^{-1} D_s \, ds\, \theta \tag{4.4}$$

and the observed Fisher information is

$$j_t^{(1)} = -\ddot{l}_t^{(1)}(\theta) = \int_0^t D_s^T c_s^{-1} D_s \, ds \tag{4.5}$$

The score vector (4.4) is a locally square integrable continuous martingale with quadratic characteristic and quadratic variation equal to $j_t^{(1)}$. Therefore, in the situation considered in this section, the three types of Fisher information, I_t, J_t and j_t, coincides as far as the information on θ is concerned. Note from (4.5) that this block of the information matrix does not depend on the parameters.

The information matrix $j_t^{(1)}$ is clearly positive semidefinite. In order to ensure that it is strictly positive definite almost surely we only need assuming that the n_1 d-dimensional random processes given by the columns of D_t are almost surely linearly independent. By this we mean the following: If, for any fixed $t > 0$, the set $\{s \le t : D_s x = 0\}$ where x is a n_1-dimensional vector has strictly positive Lebesgue measure, then $x = 0$.

Provided the columns of D_t are almost surely linearly independent, the likelihood function has almost surely a unique maximum at

$$\hat{\theta}_t = j_t^{(1)^{-1}} H_t \qquad (4.6)$$

If $\hat{\theta}_t$ belongs to the parameter set, (4.6) is the maximum likelihood estimator. In case $d = 1$ the stochastic integral in (4.3) can be expressed in terms of sums and Lebesque integrals using arguments like those given at the end of Section 3.

We can give a stronger result on the asymptotic behavior of $\hat{\theta}_t$ than those given in Section 3.

Theorem 4.1. Suppose that $i_t^{(1)}(\theta; \kappa) = E_{\theta,\kappa}(j_t^{(1)}) < \infty$ for all t, and that its diagonal elements tend to infinity as $t \to \infty$. Moreover, assume that as $t \to \infty$

$$D_t(\theta, \kappa)^{-1/2} j_t^{(1)} D_t(\theta, \kappa)^{-1/2} \to \eta^2(\theta, \kappa) \text{ in probability under } P_{\theta,\kappa} \qquad (4.7)$$

and

$$D_t(\theta, \kappa)^{-1/2} i_t^{(1)}(\theta, \kappa) D_t(\theta, \kappa)^{-1/2} \to \Sigma(\theta, \kappa) \qquad (4.8)$$

where $D_t(\theta, \kappa) = \text{diag}(i_t^{(1)}(\theta, \kappa)_{11}, \ldots, i_t^{(1)}(\theta, \kappa)_{n_1 n_1})$, $\eta^2(\theta, \kappa)$ is a random positive semidefinite matrix, and $\Sigma(\theta, \kappa)$ is a nonrandom positive definite matrix.

Then under $P_{\theta,\kappa}$ we have conditionally on $\det(\eta^2(\theta, \kappa)) > 0$:

$$\hat{\theta}_t \to \theta \quad \text{as } t \to \infty \qquad (4.9)$$

in probability, and

$$[D_t(\theta, \kappa)^{-1/2} j_t^{(1)} D_t(\theta, \kappa)^{-1/2}]^{1/2} D_t(\theta, \kappa)^{1/2} (\hat{\theta}_t - \theta) \to N(0, I_{n_1}) \qquad (4.10)$$

and

$$2(l_t^{(1)}(\hat{\theta}_t) - l_t^{(1)}(\theta)) \to \chi^2(n_1) \qquad (4.11)$$

in distribution as $t \to \infty$.

If the elements of $D_t(\theta, \kappa)$ tend to infinity at the same rate, then

$$j_t^{(1)\,1/2}(\hat{\theta}_t - \theta) \to N(0, I_{n_1}) \qquad (4.12)$$

in distribution under $P_{\theta,\kappa}$ as $t \to \infty$.

Proof. From (4.4), (4.5), and (4.6) we see that

$$j_t^{(1)}(\hat{\theta}_t - \theta) = l_t^{(1)}(\theta)$$

Since $l_t^{(1)}(\theta)$ is a continuous square integrable martingale with quadratic variation $j_t^{(1)}$, (4.10) and (4.12) follow from Theorem A.1 in the Appendix. The first conclusion is a trivial consequence of (4.10), and (4.11) follows because $2(l_t^{(1)}(\hat{\theta}_t) - l_t^{(1)}(\theta)) = (\hat{\theta}_t - \theta)^T j_t^{(1)}(\hat{\theta}_t - \theta)$.

It is worthwhile stating separately the result that the log-likelihood ratio test statistic for the hypothesis $\theta = \theta_0$ has the simple form of Wald's test statistic with the expected information replaced by j_t:

$$2(l_t^{(1)}(\hat{\theta}_t) - l_t^{(1)}(\theta_0)) = (\hat{\theta}_t - \theta_0)^T j_t^{(1)}(\hat{\theta}_t - \theta_0) \qquad (4.13)$$

We will not discuss in general under which conditions the entire likelihood function has an exponential form. Of course, this is the case under the conditions of this section when κ is fixed, that is, when the jump mechanism is known. This particular situation was considered by Skokan (1984) in the case where $d = m = k = n$, where γ is the d-dimensional identity matrix, where $\delta(t, x, z) = z$, and where, in (4.1), $a_t = 0$ and D_t is a diagonal matrix.

Conditions on Y in equation (2.13) implying an exponential structure of the part of the likelihood function that depends on κ can be found in Küchler and Sørensen (1989). A simple, but important, example is the following. Suppose the stochastic differential equation has the form (2.19). In particular, it is assumed that the jumps are generated by a compound Poisson process. We shall also assume that $k = d$, that δ does not depend on the parameters, and that $z \to \delta(t, x, z)$ has a continuously differentiable inverse. Now, if the family of jump-size distributions for the compound Poisson process is an exponential family, then the entire likelihood function has an exponential structure. Examples of this type are studied in the next section.

5 EXAMPLES

Example 5.1. Consider a stochastic differential equation of the form
(2.19) with

$$d(\theta; t, X_t) = d(t, X_t) + \gamma(t, X_t)\pi_t\theta$$

where π_t is a known nonrandom regular $d \times d$-matrix,
$d: \mathbb{R}_+ \times \mathbb{R}^d \to \mathbb{R}^d$, and where $\gamma: \mathbb{R}_+ \times \mathbb{R}^d \to \mathbb{R}^d \times \mathbb{R}^d$ is regular. As in
Section 4 we have assumed that the parameter consists of two parts θ
and κ, where the drift of the diffusion part of the process depends on
the n_1-dimensional vector θ only, while the jump mechanism depends
on the vector κ only. Moreover, we assume that $n_1 = d = m$.
     We choose $\theta^{(0)} = 0$ and find that

$$l_t^{(1)}(\theta) = \theta^T \int_0^t \pi_s^T \gamma(s, X_{s-})^{-1} \, dX_s^c(\theta^{(0)}) - \tfrac{1}{2}\theta^T \int_0^t \pi_s^T \pi_s \, ds\theta$$

see (4.2). Here

$$\int_0^t \pi_s^T \gamma(s, X_{s-})^{-1} \, dX_s^c(\theta^{(0)})$$

$$= \int_0^t \pi_s^T \gamma(s, X_{s-})^{-1} \, dX_s - \int_0^t \pi_s^T \gamma(s, X_s)^{-1} d(s, X_s) \, ds$$

$$- \sum_{s \le t} \pi_s^T \gamma(s, X_{s-}) \, \Delta X_s$$

see (2.20). It follows that

$$\hat{\theta}_t = \left[\int_0^t \pi_s^T \pi_s \, ds \right]^{-1} \int_0^t \pi_s^T \gamma(s, X_{s-})^{-1} \, dX_s^c(\theta^{(0)})$$

so that under P_θ

$$\hat{\theta}_t = \left[\int_0^t \pi_s^T \pi_s \, ds \right]^{-1} \left\{ \int_0^t \pi_s^T \gamma(s, X_{s-})^{-1} \, dX_s^c(\theta) \right.$$

$$\left. + \int_0^t \pi_s^T \gamma(s, X_{s-})^{-1}\gamma(s. X_{s-})\pi_s \, ds\theta \right\}$$

$$= \left[\int_0^t \pi_s^T \pi_s \, ds \right]^{-1} \int_0^t \pi_s^T \, dW_s + \theta$$

Now

$$\int_0^t \pi_s^T \, dW_s \sim N\left(0, \int_0^t \pi_s^T \pi_s \, ds\right)$$

so under P_θ

$$\hat{\theta}_t \sim N\left(\theta, \left[\int_0^t \pi_s^T \pi_s \, ds\right]^{-1}\right)$$

exactly. Since the log-likelihood ratio test statistic for the hypothesis $\theta = \theta_0$ is given by (4.13), we conclude that

$$2(l_t^{(1)}(\hat{\theta}_t) - l_t^{(1)}(\theta_0)) \sim \chi^2(n_1)$$

exactly, provided $\theta = \theta_0$. If the hypothesis is not true the test statistic follows a noncentral $\chi^2(n_1)$-distribution with noncentrality parameter $(\theta - \theta_0)^T (\int_0^t \pi_s^T \pi_s \, ds)(\theta - \theta_0)$.

A particular case of a stochastic differential equation of the type considered in this example is

$$dX_t = \theta X_t \, dt + \sigma X_{t-} \, dW_t + X_{t-} \, dZ_t \qquad (5.1)$$

where Z_t is a compound Poisson process

$$Z_t = \sum_{i=1}^{N_t} \epsilon_i \qquad (5.2)$$

Here N_t is a Poisson process with parameter λ while the ϵ_is are independent identically distributed random variables. The process given by (5.1) and (5.2) has been used to model the variation of stock prices, see Aase (1986a). We have to assume that σ is known. If this is not the case it can be estimated by other means, as discussed in Section 2. For convenience we will set $\sigma = 1$. Equation (5.1) has a unique solution which, by the Doléans–Dade exponential formula, is given by

$$X_t = x_o \exp[(\theta - \tfrac{1}{2})t + W_t] \prod_{i=1}^{N_t} (1 + \epsilon_i) \qquad (5.3)$$

We see that, provided $\epsilon_i > -1$, X_t is always strictly positive. Under this assumption on the distribution of ϵ_i, Condition C is satisfied for all relevant values of X_t. Here $\alpha_t(\theta)^2 c_t = \theta^2$, so the conditions of Corollary 2.2 are trivially satisfied. Hence the likelihood function exists.

The maximum likelihood estimator for θ is

$$\hat{\theta}_t = t^{-1} \int_0^t X_{s-}^{-1} \, dX_s^c(\theta^{(0)})$$

$$= t^{-1} \log\left(\frac{X_t}{x_0}\right) + \tfrac{1}{2} - t^{-1} \sum_{s \le t} \log\left(\frac{X_s}{X_{s-}}\right)$$

The latter expression for $\hat{\theta}_t$ follows by Ito's formula, see the considerations at the end of Section 3. From the general result above or by
(5.3) it follows that $\hat{\theta}_t \sim N(\theta, t^{-1})$ exactly.

The simple expression for $\hat{\theta}_t$ that we have just derived provides an
easy means of studying the effect of ignoring jumps if they are present.
If we neglect the jumps, the difference between our estimator of θ and
the correct estimator is

$$t^{-1} \sum_{i=1}^{N_t} \log(1 + \epsilon_i)$$

which is of the order $\lambda E(\log(1 + \epsilon_1))$.

Now, we will assume that the distribution of ϵ_1 belongs to a regular
one-parameter exponential family with density with respect to the Lebesgue measure. Let μ denote the mean value parameter and $T(x)$ the
canonical statistic, see Barndorff-Nielsen (1978). Since $\delta(\theta; t, x, z) = xz$, it follows from (2.18) that

$$\hat{\lambda}_t = \frac{N_t}{t} \tag{5.4}$$

and

$$\hat{\mu}_t = N_t^{-1} \sum_{s \in S_t} T\left(\frac{\Delta X_s}{X_{s-}}\right) \tag{5.5}$$

The set S_t is the set of jump times defined in Corollary 2.2. These
estimators are obviously strongly consistent and marginally asymptotically normal. The information matrices J_t and I_t are given by

$$J_t = \begin{bmatrix} t & 0 & 0 \\ 0 & \lambda^{-2}N_t & \lambda^{-1}J_t^{(1)}(\mu) \\ 0 & \lambda^{-1}J_t^{(1)}(\mu) & J_t^{(2)}(\mu) \end{bmatrix} \tag{5.6}$$

and

$$I_t = \begin{bmatrix} t & 0 & 0 \\ 0 & t\lambda^{-1} & 0 \\ 0 & 0 & t\lambda V(\mu)^{-1} \end{bmatrix} \qquad (5.7)$$

where $V(\mu) = V_\mu(T(\epsilon_1))$,

$$J_t^{(1)}(\mu) = (V(\mu))^{-1} \sum_{s \in S_t} \left(T\left(\frac{\Delta X_s}{X_{s-}}\right) - \mu \right) \qquad (5.8)$$

and

$$J_t^{(2)}(\mu) = V(\mu)^{-2} \sum_{s \in S_t} \left(T\left(\frac{\Delta X_s}{X_{s-}}\right) - \mu \right)^2 \qquad (5.9)$$

The observed Fisher information equals

$$j_t = \begin{bmatrix} t & 0 & 0 \\ 0 & \lambda^{-2} N_t & 0 \\ 0 & 0 & V(\mu)^{-1}(N_t + V'(\mu)J_t^{(1)}(\mu)) \end{bmatrix} \qquad (5.10)$$

and $i_t = I_t$.

The simultaneous normality of $(\hat{\theta}_t, \hat{\lambda}_t, \hat{\mu}_t)$ follows from Corollary 3.3. Condition (3.8) can be shown to hold provided the upper tail of the density function of ϵ_1 tends to zero quicker than x^{-3}. Remember that we have earlier assumed $\epsilon_i > -1$. From the law of large numbers it follows that $t^{-1}J_t \to \Sigma$ almost surely as $t \to \infty$, where $\Sigma = t^{-1}I_t$. Since $(\hat{\theta}_t, \hat{\lambda}_t, \hat{\mu}_t)$ is strongly consistent, so is the sequence $(\tilde{\theta}_t, \tilde{\lambda}_t, \tilde{\mu}_t)$ in condition (3.15). Now it follows from the law of large numbers that $t^{-1}j_t(\tilde{\theta}_t, \tilde{\lambda}_t, \tilde{\mu}_t) \to \Sigma$ almost surely as $t \to \infty$. Hence (3.14) and (3.19) follow.

Example 5.2. The stochastic differential equation

$$dX_t = (\theta_1 + \theta_2 X_t)\, dt + \sigma\, dW_t + dZ_t \qquad (5.11)$$

where Z_t is a compound Poisson process, has been used to model the dynamics of soil moisture, see Mtundu and Koch (1987). Let Z_t be defined as in Example 5.1, however without restriction on the possible values of ϵ_i. The statistical problem is to estimate $(\theta_1, \theta_2, \lambda, \mu)$. For the dominating measure we choose $(\theta_1, \theta_2, \lambda, \mu) = (0, 0, 1, \mu^{(0)})$.

The equation (5.11) has a unique solution given by

$$X_t = x_0 e^{\theta_2 t} + \frac{\theta_1}{\theta_2} (e^{\theta_2 t} - 1) + \sigma \int_0^t e^{\theta_2(t-s)} \, dW_s$$

$$+ \int_0^t e^{\theta_2(t-s)} \, dZ_s \qquad (5.12)$$

provided $\theta_2 \neq 0$, and by

$$X_t = \theta_1 t + \sigma W_t + Z_t \qquad (5.13)$$

when $\theta_2 = 0$. The result (5.12) is proved by means of Ito's formula. Condition C is clearly satisfied, and since $\alpha_t(\theta)^2 c_t = (\theta_1 + \theta_2 X_t)^2$ the conditions of Corollary 2.2 are met too. Hence the likelihood function exists.

By the results of Section 4

$$\hat{\theta}_{1t} = \frac{X_t^c(\theta_1^{(0)}, \theta_2^{(0)}) \int_0^t X_s^2 \, ds - \int_0^t X_s \, ds \int_0^t X_{s-} \, dX_s^c(\theta_1^{(0)}, \theta_2^{(0)})}{t \int_0^t X_s^2 \, ds - (\int_0^t X_s \, ds)^2} \qquad (5.14)$$

and

$$\hat{\theta}_{2t} = \frac{t \int_0^t X_{s-} \, dX_s^c(\theta_1^{(0)}, \theta_2^{(0)}) - X_t^c(\theta_1^{(0)}, \theta_2^{(0)}) \int_0^t X_s \, ds}{t \int_0^t X_s^2 \, ds - (\int_0^t X_s \, ds)^2} \qquad (5.15)$$

where

$$X_t^c(\theta_1^{(0)}, \theta_2^{(0)}) = X_t - \sum_{s \le t} \Delta X_s$$

and by Ito's formula

$$\int_0^t X_{s-} \, dX_s^c(\theta_1^{(0)}, \theta_2^{(0)}) = \tfrac{1}{2} \left[X_t^2 - x_0^2 - \sigma^2 t - \sum_{s \le t} (X_s^2 - X_{s-}^2) \right]$$

The denominator in (5.14) and (5.15) is strictly positive unless X_s is constant on $[0, t]$, which happens with probability zero (for $\sigma \neq 0$). The maximum likelihood estimates $\hat{\lambda}_t$ and $\hat{\mu}_t$ are given by (5.4) and (5.5) with $\Delta X_s / X_{s-}$ replaced by ΔX_s. The information matrix J_t equals

$$\begin{bmatrix} \sigma^{-2} t & \sigma^{-2} \int_0^t X_s \, ds & 0 & 0 \\ \sigma^{-2} \int_0^t X_s \, ds & \sigma^{-2} \int_0^t X_s^2 \, ds & 0 & 0 \\ 0 & 0 & \lambda^{-2} N_t & \lambda^{-1} J_t^{(1)}(\mu) \\ 0 & 0 & \lambda^{-1} J_t^{(1)}(\mu) & J_t^{(2)}(\mu) \end{bmatrix}$$

where $J_t^{(1)}(\mu)$ and $J_t^{(2)}(\mu)$ are defined by (5.8) and (5.9), again with $\Delta X_s / X_{s-}$ replaced by ΔX_s.

In the following considerations we set $\sigma^2 = 1$ for convenience. From (5.12) it follows that for $\theta_2 \neq 0$

$$E(X_t) = (x_0 + \kappa)e^{\theta_2 t} - \kappa \tag{5.16}$$

and

$$E(X_t^2) = ((x_0 + \kappa)^2 + \nu)e^{2\theta_2 t} - 2\kappa(x_0 + \kappa)e^{\theta_2 t} + (\kappa^2 - \nu) \tag{5.17}$$

where $\kappa = (\theta_1 + \lambda m)/\theta_2$, $\nu = \frac{1}{2}(1 + \lambda m^{(2)})/\theta_2$, $m = E_\mu(\epsilon_1)$ and $m^{(2)} = E_\mu(\epsilon_1^2)$. The equations (5.16) and (5.17) follow easily using that $Z_t - \lambda mt$ is a martingale with quadratic characteristic $\lambda m^{(2)}t$. From (5.16) and (5.17) we conclude that

$$E\left(\int_0^t X_s \, ds\right) = \theta_2^{-1}(x_0 + \kappa)(e^{\theta_2 t} - 1) - \kappa t \tag{5.18}$$

and

$$E\left(\int_0^t X_s^2 \, ds\right) = \frac{1}{2}\theta_2^{-1}((x_0 + \kappa)^2 + \nu)(e^{2\theta_2 t} - 1)$$
$$- 2\theta_2^{-1}\kappa(x_0 + \kappa)(e^{\theta_2 t} - 1) + (\kappa^2 - \nu)t \tag{5.19}$$

When $\theta_2 = 0$ we find from (5.13) that

$$E\left(\int_0^t X_s \, ds\right) = \frac{1}{2}(\theta_1 + \lambda m)t^2 \tag{5.20}$$

and

$$E\left(\int_0^t X_s^2 \, ds\right) = \frac{1}{2}(1 + \lambda m^{(2)})t^2 + \frac{1}{3}(\theta_1 + \lambda m)^2 t^3 \tag{5.21}$$

The upper left corner of the expected Fisher information matrix i_t can be found using (5.18)–(5.21), while the lower right corner is as in (5.7).

Let us first consider the case $\theta_2 < 0$. We see that $t^{-1}i_t \rightarrow \Sigma$ as $t \rightarrow \infty$, where

$$\Sigma = \begin{bmatrix} 1 & -\kappa & 0 & 0 \\ -\kappa & (\kappa^2 - \nu) & 0 & 0 \\ 0 & 0 & \lambda^{-1} & 0 \\ 0 & 0 & 0 & \lambda V(\mu)^{-1} \end{bmatrix}$$

with $V(\mu)$ defined as in the previous example. Since $\nu < 0$ for $\theta_2 < 0$, the matrix Σ is positive definite. By an ergodic theorem

$$t^{-1} \int_0^t X_s \, ds \to -\kappa \qquad \text{as } t \to \infty \tag{5.22}$$

and

$$t^{-1} \int_0^t X_s^2 \, ds \to \kappa^2 - \nu \qquad \text{as } t \to \infty \tag{5.23}$$

almost surely when $\theta_2 < 0$. This is because the autocorrelation decreases exponentially. From (5.22), (5.23) and the law of large numbers it follows that $t^{-1}J_t \to \Sigma$ almost surely. Condition (3.8) can be verified when the tails of the density function of ϵ_1 tend to zero quicker than $|x|^{-3}$. The observed information matrix consists of two blocks: the upper left block equals that of J_t, while the lower right block equals the (λ, μ)-part of (5.10) with $J_t^{(1)}(\mu)$ redefined as indicated above. Thus $j_t(\theta_1, \theta_2, \lambda, \mu)$ does not depend on (θ_1, θ_2) and condition (3.15) can be checked like in Example 5.1. Now it follows from Theorem 3.1 and Corollary 3.3 that the score function and the maximum likelihood estimator are asymptotically normal when properly normalized. It also follows that $\hat{\theta}_{1t}$ and $\hat{\theta}_{2t}$ are consistent. The following argument shows that they are indeed strongly consistent.

$$\hat{\theta}_{1t} - \theta_1$$
$$= \frac{X_t^c(\theta_1, \theta_2) \int_0^t X_s^2 \, ds - \int_0^t X_s \, ds \int_0^t X_{s-} \, dX_s^c(\theta_1, \theta_2)}{t \int_0^t X_s^2 \, ds - (\int_0^t X_s \, ds)^2}$$
$$= \frac{t^{-1}X_t^c(\theta_1, \theta_2) t^{-1} \int_0^t X_s^2 \, ds - t^{-1} \int_0^t X_s \, ds \, t^{-1} \int_0^t X_{s-} \, dX_s^c(\theta_1, \theta_2)}{t^{-1} \int_0^t X_s^2 \, ds - (t^{-1} \int_0^t X_s \, ds)^2}$$

$$\tag{5.24}$$

It follows from (5.22), (5.23), and Lepingle's (1978) law of large numbers that this expression converges almost surely to zero because $X_t^c(\theta_1, \theta_2)$ is a martingale with quadratic characteristic t, and $\int_0^t X_{s-} \, dX_s^c(\theta_1, \theta_2)$ is a martingale with quadratic characteristic $\int_0^t X_s^2 \, ds \sim t$. The strong consistency of $\hat{\theta}_{2t}$ is proved in the same way using that

$$\hat{\theta}_{2t} - \theta_2 = \frac{t \int_0^t X_{s-} \, dX_s^c(\theta_1, \theta_2) - X_t^c(\theta_1, \theta_2) \int_0^t X_s \, ds}{t \int_0^t X_s^2 \, ds - (\int_0^t X_s \, ds)^2} \tag{5.25}$$

When $\theta_2 > 0$ the components of i_t increase to infinity at different rates. Therefore we use Corollary 3.2 to deduce simultaneous asymptotic normality of the maximum likelihood estimators. In Corollary 3.2 we can, for convenience, set $D_t(\theta) = \mathrm{diag}(t, e^{2\theta_2 t}, t, t)$. First we consider the conditions (3.9) and (3.15).

From (5.12) we see that

$$
\begin{aligned}
H_t &= e^{-\theta_2 t} X_t - x_0 - \kappa(1 - e^{-\theta_2 t}) \\
&= \int_0^t e^{-\theta_2 s} \, dW_s + \int_0^t e^{-\theta_2 s} \, d\tilde{Z}_s
\end{aligned}
$$

is a martingale. Here $\tilde{Z}_t = Z_t - \lambda m t$. Since $E(H_t^2) = \nu(1 - e^{-2\theta_2 t}) \leq \nu$ it follows from the martingale convergence theorem that $e^{-\theta_2 t} X_t$ converges almost surely to a random variable $H_\infty + x_0 + \kappa$. Hence $e^{-\theta_2 t} X_t^2$ converges almost surely to $(H_\infty + x_0 + \kappa)^2$, and by the integral version of the Toeplitz lemma

$$
e^{-\theta_2 t} \int_0^t X_s \, ds \rightarrow \frac{H_\infty + x_0 + \kappa}{\theta_2} \tag{5.26}
$$

and

$$
e^{-2\theta_2 t} \int_0^t X_s^2 \, ds \rightarrow \frac{\tfrac{1}{2}(H_\infty + x_0 + \kappa)^2}{\theta_2} \tag{5.27}
$$

almost surely as $t \to \infty$. From these results and earlier considerations (3.9) and (3.15) follow with $\eta^2 = \mathrm{diag}(1, \tfrac{1}{2}(H_\infty + x_0 + \kappa)^2/\theta_2, \lambda^{-1}, \lambda V(\mu)^{-1})$. To establish the simultaneous asymptotic normality of the maximum likelihood estimator and of the score vector we only need to check (3.8) and (3.10). As earlier (3.8) is satisfied if the tails of the density of ϵ_1 tend to infinity faster than $|x|^{-3}$, and (3.10) is obviously satisfied with $\Sigma = \mathrm{diag}(1, \tfrac{1}{2}((x_0 + \kappa)^2 + \nu)/\theta_2, \lambda^{-1}, \lambda V(\mu)^{-1})$. The strong consistency of $(\hat{\theta}_{1t}, \hat{\theta}_{2t})$ follows in a way analogous to that in the case $\theta_2 < 0$.

Example 5.3. Consider the stochastic differential equation

$$
dX_t = \theta_1 X_t(1 - X_t/K) \, dt - X_{t-} \, dZ_t + \sigma X_{t-} \, dW_t \tag{5.28}
$$

where Z_t is a compound Poisson process defined as in the previous examples. Here we will make the extra assumption that the probability distribution of the jumps of Z_t are concentrated in the interval $(0, 1)$.

Equation (5.28) models the dynamics of a population that grows logistically between disasters. A disaster, when it occurs, removes a random fraction of the population. The parameter K is the carrying capacity of the environment. The model proposed here generalizes the model of Hanson and Tuckwell (1981). It is also a generalization of a diffusion population model studied by Guttorp and Kulperger (1984).

Equation (5.28) has a unique solution. This follows from Theorem III-2-32 of Jacod and Shiryaev (1987). Using the Doléans-Dade exponential formula, we find that

$$X_t = x_0 \exp\left((\theta_1 - \tfrac{1}{2}\sigma^2)t - \theta_1 K^{-1} \int_0^t X_s \, ds + \sigma W_t \right.$$

$$\left. + \sum_{i=1}^{N_t} \log(1 - \epsilon_i) \right) \tag{5.29}$$

Hence, X_t is always positive provided $x_0 > 0$. The biologically interesting problem is to find the probability that X_t in a finite time interval hits a small critical level at which the population is virtually extinct. From (5.29) we see that $X_t \le x_0 \exp((\theta_1 - \tfrac{1}{2}\sigma^2)t + \sigma W_t)$, and hence that if $\theta_1 < \tfrac{1}{2}\sigma^2$ the population size will tend to zero almost surely as $t \to \infty$. We note that if σ^2 is estimated without taking the jumps into account, it will be overestimated, see the discussion in Section 2. As a consequence erroneous conclusions about the probability of extinction can be made. We will not here go into the case where $\theta_1 \ge \tfrac{1}{2}\sigma^2$, but only consider estimation of the parameters of the model.

We introduce the parameter $\theta_2 = -\theta_1 K^{-1}$. To be specific we assume that the distribution of ϵ_1 is a beta-distribution on $(0, 1)$ with density function $x^{\kappa_1 - 1}(1 - x)^{\kappa_2 - 1} B(\kappa_1, \kappa_2)^{-1}$. Condition C is satisfied for all relevant values of X_t, and the conditions of Corollary 2.2 hold because $\alpha_t(\theta)^2 c_t = (\theta_1 + \theta_2 X_t)^2$. Hence the likelihood function exists. For the dominating measure we have used $(\theta_1, \theta_2, \lambda, \kappa_1, \kappa_2) = (0, 0, 1, 1, 1)$. By the results of Section 4

$$\hat{\theta}_{1t} = \frac{\int_0^t X_{s-}^{-1} \, dX_s^c(\theta_1^{(0)}, \theta_2^{(0)}) \int_0^t X_s^2 \, ds - X_t^c(\theta_1^{(0)}, \theta_2^{(0)}) \int_0^t X_s \, ds}{t \int_0^t X_s^2 \, ds - (\int_0^t X_s \, ds)^2}$$

and

$$\hat{\theta}_{2t} = \frac{t X_t^c(\theta_1^{(0)}, \theta_2^{(0)}) - \int_0^t X_{s-}^{-1} \, dX_s^c(\theta_1^{(0)}, \theta_2^{(0)}) \int_0^t X_s \, ds}{t \int_0^t X_s^2 \, ds - (\int_0^t X_s \, ds)^2}$$

where

$$X_t^c(\theta_1^{(0)}, \theta_2^{(0)}) = X_t - \sum_{s \le t} \Delta X_s$$

and by Ito's formula

$$\int_0^t X_{s-}^{-1}\, dX_s^c(\theta_1^{(0)}, \theta_2^{(0)}) = \log(X_t/x_0) + \tfrac{1}{2}\sigma^2 t - \sum_{s \le t} \log(X_s/X_{s-})$$

It is easily found that $\hat{\lambda}_t$ is given by (5.4), and that $\hat{\kappa}_{1t}$ and $\hat{\kappa}_{2t}$ solve

$$\psi(\hat{\kappa}_1) - \psi(\hat{\kappa}_1 + \hat{\kappa}_2) = N_t^{-1} \sum_{s \le t} \log(-\Delta X_s/X_{s-})$$

and

$$\psi(\hat{\kappa}_2) - \psi(\hat{\kappa}_1 + \hat{\kappa}_2) = N_t^{-1} \sum_{s \le t} \log(X_s/X_{s-})$$

where ψ is the digamma function.

For the particular model considered here we can give another estimate of σ^2 than the one proposed in Section 2. From (5.29) it follows that as $n \to \infty$

$$t^{-1}\left\{\sum_{j=1}^{2^n} (R_{jt2^{-n}} - R_{(j-1)t2^{-n}})^2 - \sum_{s \le t} (\Delta R_s)^2\right\} \to \sigma^2$$

in probability, where $R_t = \log(X_t)$.

Example 5.4. Consider a one-dimensional stochastic differential equation of the type (2.3) with

$$\beta(\tilde{\theta}_1, \theta_2; t, x) = \tilde{\theta}_1 + \theta_2 x$$

$$\gamma = 1$$

$$\delta(\theta; t, x, z) = z$$

and

$$q^{\alpha}(dt, dz) = z^{-1} e^{-\alpha z}\, dz\, dt$$

Here $(\tilde{\theta}_1, \theta_2, \alpha) \in \mathbb{R} \times \mathbb{R} \times \mathbb{R}_+$. The measure q^{α} is the product of the Lévy measure of the gamma-process and the Lebesgue measure. This equation has a unique solution by Theorem III-2-32 in Jacod and Shiryaev (1987). Condition C is clearly satisfied since $b(\tilde{\theta}_1, \theta_2, \alpha; t, x) = \tilde{\theta}_1 - \alpha^{-1} e^{-\alpha} + \theta_2 x$.

We will use the parameterization $(\theta_1, \theta_2, \alpha)$, where $\theta_1 = \tilde{\theta}_1 - \alpha^{-1}$ is the part of the nonrandom drift that is not due to the jumps. For the dominating measure in the likelihood function we choose $(0, 0, 1)$. With that choice

$$Y(\alpha; t, x, y) = e^{(1-\alpha)y}$$

and

$$y_t(\theta) = \theta_1 + \theta_2 X_{t-} + 1$$

Now, since $e^{(1-\alpha)y} - 1 \sim y$ as $y \to 0$, we see that the integrals

$$\int_0^\infty (e^{(1-\alpha)y} - 1)y^{-1}e^{-y}\,dy$$

and

$$\int_0^\infty (e^{(1-\alpha)y} - 1)^2 1_{\{y \le 2\}} y^{-1} e^{-y}\,dy$$

converge. Hence the conditions of Theorem 2.1 are satisfied and the likelihood function exists. The model considered here is of the type studied in Section 4. Therefore we see immediately that $(\hat{\theta}_{1t}, \hat{\theta}_{2t})$ are given by (5.14) and (5.15). The strong consistency of these estimators can be proved in the same way as in Example 5.2. The asymptotic normality is verified in a similar way. The second part of the likelihood function is

$$l_t^{(2)}(\alpha) = \int_0^t \int_0^\infty [e^{(1-\alpha)y} - 1](\mu - \nu(1))(dy, ds)$$

$$+ \int_0^t \int_0^\infty [-(\alpha - 1)y - e^{(1-\alpha)y} + 1]\mu(dy, ds)$$

$$= -t \int_0^\infty [e^{-\alpha y} - e^{-y}]y^{-1}\,dy$$

$$- (\alpha - 1) \int_0^t \int_0^\infty y\mu(dy, ds)$$

$$= t \log(\alpha) - (\alpha - 1)Z_t$$

where $Z_t = \int_0^t \int_0^\infty y\mu(dy, ds)$, the sum of the jumps, is a gamma process. Note that the model is an exponential family. We see that

$$\hat{\alpha}_t^{-1} = t^{-1}Z_t$$

By standard results for processes with independent stationary increments (or for exponential families)

$$\hat{\alpha}_t \to \alpha \qquad \text{as } t \to \infty$$

almost surely, and

$$\alpha\sqrt{t}(\hat{\alpha}_t^{-1} - \alpha^{-1}) = \alpha(Z_t - \alpha^{-1}t)/\sqrt{t} \to N(0, 1)$$

in distribution as $t \to \infty$, since $\text{Var}(Z_t) = t\alpha^{-2}$. Hence

$$\alpha^{-1}\sqrt{t}(\hat{\alpha}_t - \alpha) \to N(0, 1)$$

in distribution.

APPENDIX: A CENTRAL LIMIT THEOREM FOR MULTIVARIATE MARTINGALES

The processes considered in this Appendix are supposed to be cadlag and defined on a complete filtered probability space $(\Omega, \mathscr{F}, \{\mathscr{F}_t\}, P)$ satisfying the usual conditions. We denote by I_n the $n \times n$ identity matrix, by $\det(A)$ the determinant of an $n \times n$ matrix, and by $\text{diag}(x_1, \ldots, x_n)$ the $n \times n$ diagonal matrix with $x_1, \ldots, x_n$ in the diagonal. A superscript T denotes transposition, while $A^{1/2}$ denotes the unique positive semidefinite square root of a positive semidefinite matrix A. The concepts of stable and mixing convergence used in the theorem were introduced by Rényi (1958, 1963) and further developed by Aldous and Eagleson (1978), see also the discussion in Hall and Heyde (1980).

Theorem A.1. Let $M = (M_1, \ldots, M_n)^T$ be an n-dimensional square integrable $\{\mathscr{F}_t\}$-martingale with quadratic variation matrix $[M]$, and set $H_t = E(M_t M_t^T)$. Suppose there exists a nonrandom vector function $k_t = (k_{1t}, \ldots, k_{nt})^T$ with $k_{it} > 0$ increasing to infinity as $t \to \infty$ for $i = 1, \ldots, n$ and such that the following holds as $t \to \infty$:

1. $k_{it}^{-1}E(\sup_{s \le t}|\Delta M_{is}|) \to 0$ $i = 1, \ldots, n$,
2. $K_t^{-1}[M]_t K_t^{-1} \to \eta^2$ in probability,
   where $K_t = \text{diag}(k_{1t}, \ldots, k_{nt})$ and η^2 is a random positive semidefinite matrix,
3. $K_t^{-1}H_t K_t^{-1} \to \Sigma$,
   where Σ is a positive definite matrix.

Then we have the following results about convergence in distribution:

$$K_t^{-1}M_t \to Z \quad \text{(stably)} \tag{A.1}$$

where the distribution of Z is the normal variance mixture with characteristic function $\varphi(u) = E(\exp(-\frac{1}{2}u^T\eta^2 u))$, $u = (u_1, \ldots, u_n)^T$,

$$(K_t[M]_t^{-1}K_t)^{1/2}K_t^{-1}M_t | \{\det(\eta^2) > 0\} \rightarrow N(0, I_n) \quad \text{(mixing)} \quad \text{(A.2)}$$

and

$$M_t^T[M]_t^{-1}M_t | \{\det(\eta^2) > 0\} \rightarrow \chi^2(n) \quad \text{(mixing)} \quad \text{(A.3)}$$

Proof. To prove (A.1) it suffices to show that $x^T K_t^{-1}M_t \rightarrow x^T Z$ (stably) for all $x \in \mathbb{R}^n$ (the Cramér-Wold device). For every $t > 0$ the process

$$X_s^t = (x^T K_t^{-1}H_t K_t^{-1}x)^{-1/2}x^T K_t^{-1}M_{st} \quad 0 \leq s \leq 1$$

is a one-dimensional square integrable martingale with respect to the filtration $\{\mathscr{F}_{st}\}_{s \in [0,1]}$. Using (1), (2) and (3) we see that, as $t \rightarrow \infty$,

$$E(\sup_{0 \leq s \leq 1} |\Delta X_s^t|) \leq (x^T K_t^{-1}H_t K_t^{-1}x)^{-1/2}$$
$$\times \sum_i |x_i| k_{it}^{-1} E(\sup_{s \leq t} |\Delta M_{is}|) \rightarrow 0$$

and

$$[X^t]_1 = (x^T K_t^{-1}H_t K_t^{-1}x)^{-1}x^T K_t^{-1}[M]_t K_t^{-1}x \rightarrow (x^T\Sigma x)^{-1}x^T\eta^2 x$$

in probability. Hence the class of processes X^t, $t > 0$, satisfies the same conditions as the martingales (3.6) in the proof of Theorem 2 in Feigin (1985). Now it follows directly from Feigin's proof that X_1^t converges stably in distribution to the zero-mean normal variance mixture with mixing distribution $(x^T\Sigma x)^{-1}x^T\eta^2 x$. From this conclusion $x^T K_t^{-1}M_t \rightarrow x^T Z$ (stably), and hence (A.1), follows immediately.

The stability of (A.1) implies that conditionally on $\det(\eta^2) > 0$

$$\eta^{-1}K_t^{-1}M_t \rightarrow N(0, I_n) \quad \text{(stably)}$$

and from this (A.2) and (A.3) follow under condition (2).

Theorem A.1 generalizes a theorem by Hutton and Nelson (1984b). Their theorem, where $k_{1t} = \cdots = k_{nt}$, does not cover cases where the variance of the components of M increase at different rates. This can easily happen in statistical applications (see Example 5.2). Usually we can set $k_{it} = \text{Var}(M_{is})^{1/2}$.

REFERENCES

Aase, K. K. (1986a). New option pricing formulas when the stock model is a combined continuous/point process. *Technical Report No. 60*, CCREMS, MIT, Cambridge, MA.

Aase, K. K. (1986b). Probabilistic solutions of option pricing and corporate security valuation. *Technical Report No. 63*, CCREMS, MIT, Cambridge, MA.

Aase, K. K., and Guttorp, P. (1987). Estimation in models for security prices. *Scand. Actuarial J.*, *3–4*, 211–224.

Aldous, D. J., and Eagleson, G. K. (1978). On mixing and stability of limit theorems. *Ann. Probability*, *6*, 325–331.

Barndorff-Nielsen, O. E. (1978). *Information and Exponential Families*. Wiley Chichester.

Barndorff-Nielsen, O. E., and Sørensen, M. (1989). Asymptotic likelihood theory for stochastic processes. A review. *Research Report No. 162*, Department of Theoretical Statistics, Aarhus University. To appear in *Int. Statist. Rev.*

Basawa, I. V., and Prakasa Rao, B. L. S. (1980). *Statistical Inference for Stochastic Processes*. Academic Press, London.

Bodo, B. A., Thompson, M. E., and Unny, T. E. (1987). A review on stochastic differential equations for applications in hydrology. *Stochastic Hydrol. Hydraul.*, *1*, 81–100.

Feigin, P. (1985). Stable convergence of semimartingales. *Stoc. Proc. Appl.*, *19*, 125–134.

Godambe, V. P., and Heyde, C. C. (1987): Quasi-likelihood and optimal estimation. *Int. Statist. Rev.*, *55*, 231–244.

Guttorp, P., and Kulperger, R. (1984). Statistical inference for some Volterra population processes in a random environment. *Can. J. Statist.*, *12*, 289–302.

Hall, P., and Heyde, C. C. (1980). *Martingale Limit Theory and Its Application*. Academic Press, New York.

Hanson, F. B., and Tuckwell, H. C. (1981). Logistic growth with random density independent disasters. *Theor. Pop. Biol.*, *19*, 1–18.

Hutton, J. E., and Nelson, P. I. (1984a). Interchanging the order of differentiation and stochastic integration. *Stoc. Proc. Appl.*, *18*, 371–377.

Hutton, J. E., and Nelson, P. I. (1984b). A mixing and stable cental limit theorem for continuous time martingales. *Technical Report No. 42*, Kansas State University.

Hutton, J. E., and Nelson, P. I. (1986). Quasi-likelihood estimation for semi-martingales. *Stoc. Proc. Appl.*, *22*, 245–257.

Jacod, J. (1979). *Calcul Stochastique et Problèmes de Martingales.* Lecture Notes in Mathematics Vol. 714, Springer, Berlin-Heidelberg-New York.

Jacod, J., and Mémin, J. (1976). Caractéristiques locales et conditions de continuité absolue pour les semi-martingales. *Z. Wahrscheinlichkeitstheorie verw. Gebiete, 35,* 1–37.

Jacod, J., and Shiryaev, A. N. (1987). *Limit Theorems for Stochastic Processes.* Springer, New York.

Karandikar, R. L. (1983). Interchanging the order of stochastic integration and ordinary differentiation. *Sankhya, Ser. A, 45,* 120–124.

Küchler, U., and Sørensen, M. (1989). Exponential families of stochastic processes: A unifying semimartingale approach. *Int. Statist. Rev., 57,* 123–144.

Lepingle, D. (1978). Sur le comportement asymptotique des martingales locales. In Dellacherie, C., Meyer, P. A. and Weil, M. (eds.), *Séminaire de Probabilités XII, Lecture Notes in Mathematics, 649.* Springer, Berlin-Heidelberg-New York.

Lin'kov, Ju. N. (1982). Asymptotic properties of statistical estimators and tests for Markov processes. *Theor. Probability and Math. Statist., 25,* 83–98.

Lin'kov, Ju. N. (1983). Asymptotic normality of locally square-integrable martingales in a series scheme. *Theor. Probability and Math. Statist., 27,* 95–103.

Liptser, R. S., and Shiryaev, A. N. (1977). *Statistics of Random Processes.* Springer, New York.

Mtundu, N. D., and Koch, R. W. (1987). A stochastic differential equation approach to soil moisture. *Stochastic Hydrol. Hydraul., 1,* 101–116.

Rényi, A. (1958). On mixing sequences of sets. *Acta Math. Acad. Sci. Hung., 9,* 215–228.

Rényi, A. (1963). On stable sequences of events. *Sankhya, Ser. A, 25,* 293–302.

Skokan, V. (1984). A contribution to maximum likelihood estimation in processes given by certain types of stochastic differential equations. In P. Mandl and M. Hušková (eds.), *Asymptotic Statistics, Vol. 2.* Elsevier, Amsterdam-New York-Oxford.

Sørensen, M. (1983). On maximum likelihood estimation in randomly stopped diffusion-type processes. *Int. Statist. Rev., 51,* 93–110.

Sørensen, M. (1989a). A note on the existence of a consistent maximum likelihood estimator for diffusions with jumps. In Langer, H. and Nollau,

V. (eds.), *Markov Processes and Control Theory*, pp. 229–234. Akademie-Verlag, Berlin.

Sørensen, M. (1989b). Some asymptotic properties of quasi likelihood estimators for semimartingales. In Mandel, P., and Hušková, M. (eds.), *Proceedings of the Fourth Prague Symposium on Asymptotic Statistics*, pp. 469–479. Charles University, Prague.

Sørensen, M. (1990). On quasi likelihood for semimartingales. *Stoch. Processes Appl.*, *35*, 331–346.

Stroock, D. W. (1975). Diffusion processes associated with Lévy generators. *Z. Wahrscheinlichkeitstheorie verw. Gebiete*, *32*, 209–244.

4

Efficient Estimating Equations for Nonparametric Filtered Models

P. E. Greenwood*
University of British Columbia, Vancouver, British Columbia, Canada

W. Wefelmeyer†
University of Cologne, Cologne, Federal Republic of Germany

Let $X_{n1}, \ldots, X_{nn}$ be counting processes with intensity processes known up to a (possibly infinite-dimensional) parameter. The (partially specified) likelihood admits a representation in terms of the intensity processes. We give conditions for uniform local asymptotic normality and consider the problem of constructing efficient estimators for functionals of the parameter. The standardized error of a regular and asymptotically efficient estimator is stochastically approximable by the derivative of the likelihood in the direction least favorable for the functional. We show how this characterization can be used to prove efficiency for (i) certain martingale estimators, (ii) an infinite-dimensional version of the Newton–Raphson improvement of a preliminary estimator, (iii) solutions of estimating equations based on derivatives of the likelihood, including nonparametric maximum likelihood estimators (in the sense of Gill, 1989).

The results are illustrated by estimators for (i) the cumulative hazard

---

*Work supported by an NSF Visiting Professorship and NSERC, Canada.
†Work supported by a Feodor Lynen grant of the Alexander von Humboldt Foundation, and NSERC, Canada.

rate in Aalen's additive risk model, (ii) the vector of regression coefficients in Andersen and Gill's counting process version of Cox's proportional hazards model.

1 INTRODUCTION

Let $(\Omega_n, \mathcal{F}_n, \mathbb{G}_n, P_{n\vartheta} : \vartheta \in \Theta)$ be a filtered model with arbitrary (possibly infinite-dimensional) parameter space Θ. Suppose we observe counting processes $X_{n1}, \ldots, X_{nn}$ and covariate processes $Y_{n1}, \ldots, Y_{nn}$ on $[0, 1]$, say. Assume that the Doob–Meyer decomposition of X_{ni} with respect to $\mathbb{G}_n$ and $P_{n\vartheta}$ is of the form

$$X_{ni}(t) = M_{ni\vartheta}(t) + \int_0^t a_{ni\vartheta}(s, Y_{ni}(s))\, ds$$

with $M_{ni\vartheta}$ a martingale and $a_{ni\vartheta}(s, Y_{ni}(s))$ a predictable *intensity process*.

Various methods have been proposed for constructing asymptotically efficient estimators for semiparametric models based on i.i.d. observations. Our purpose is to unify and extend these to stochastic process models with infinite-dimensional parameter, including semiparametric models.

We restrict ourselves to *counting processes* X_{ni}. Other types of semimartingales can be treated similarly. We do not assume that the covariate processes are adapted to the filtration generated by the counting processes. In this case, the distribution of the counting process is not determined by the intensity process. This means that the model is only *partially specified* in the sense of Arjas and Haara (1984) and Greenwood (1988). We include partially specified models in order to cover counting process regression models of interest in survival analysis (Sections 7 and 8). Fortunately, this does not lead to additional technical complications. It was shown in Greenwood and Wefelmeyer (1990a) that an asymptotic efficiency concept can be based on the *partially specified log-likelihood* between τ and ϑ,

$$L_{n\vartheta\tau} = \sum_{i=1}^{n} \left(\int_0^1 \log(a_{ni\tau}(s, Y_{ni}(s))/a_{ni\vartheta}(s, Y_{ni}(s)))\, dX_{ni}(s) \right.$$

$$\left. - \int_0^1 (a_{ni\tau}(s, Y_{ni}(s)) - a_{ni\vartheta}(s, Y_{ni}(s)))\, ds \right) \qquad (1.1)$$

(Set $\log(\infty) = 0$.) If the filtration $\mathbb{G}_n$ equals the filtration generated by the counting processes, no two of which jump simultaneously, this is the usual representation of the full likelihood. [For multivariate point processes, this representation was introduced by Jacod, 1975, p. 250, Theorem (5.1); see also Jacod and Shiryaev, 1987, p. 190, Theorem 5.45.] For the counting process setting considered here, the partially specified likelihood was introduced by Gill (1985) and Slud (1986, Chapter 6). A corresponding concept for general semimartingales was introduced by Jacod (1987) and (1990).

Asymptotic efficiency is understood in the sense of a Hájek–Le Cam convolution theorem. An appropriate local asymptotic normality is formulated in Section 3. Following Le Cam (1986, p. 179, Section 10.3), we introduce a *local parameterization*

$$\xi_{n\vartheta}: \Theta \to H_\vartheta$$

The map $\xi_{n\vartheta}$ will rescale $\tau \in \Theta$ around ϑ as $c_n(\tau - \vartheta)$ with $c_n \to \infty$ and then apply an approximate linear isometry from Θ into some Hilbert space, H_ϑ, with norm $\| \ \|_\vartheta$ determined by the structure of the model.

Assume that the intensity processes $a_{ni\,\vartheta}$ are *Hellinger differentiable* at ϑ with random *differential operator* $D_{ni\,\vartheta}$, and that the *Hellinger sequence* is *asymptotically nonrandom*. (See conditions (3.3) to (3.5).) Then the (partially specified) log-likelihood is *locally asymptotically normal* for τ near ϑ,

$$L_{n\vartheta\tau} = c_n^{-1} \sum_{i=1}^n \int_0^1 D_{ni\,\vartheta}(\xi_{n\vartheta}(\tau))(s) \, dM_{ni\,\vartheta}(s)$$

$$-\tfrac{1}{2}\|\xi_{n\vartheta}(\tau)\|_\vartheta^2 + o_{P_{n\vartheta}}(1) \tag{1.2}$$

the martingale term is asymptotically normal with variance $\|\xi_{n\vartheta}(\tau)\|_\vartheta^2$. The variance determines how difficult it is, asymptotically, to distinguish between ϑ and τ. We call the corresponding inner product $(\ , \ )_\vartheta$ the *acuity*.

Let us first consider the construction of efficient estimators for a real-valued functional

$$k: \Theta \to \mathbb{R}$$

We will see below that estimators for infinite-dimensional parameters can be obtained from this case.

Assume that the functional is *differentiable* at ϑ with *gradient* $g_\vartheta \in H_\vartheta$ in the sense that, for τ near ϑ,

$$c_n(k(\tau) - k(\vartheta)) - (\xi_{n\vartheta}(\tau), g_\vartheta)_\vartheta \to 0 \qquad (1.3)$$

The gradient is not uniquely determined. Any function f_ϑ with

$$(\xi_{n\vartheta}(\tau), f_\vartheta - g_\vartheta)_\vartheta \to 0$$

is also a gradient. Assume that g_ϑ is chosen *canonical* in the sense that it is asymptotically tangent to $\xi_{n\vartheta}(\Theta)$.

According to Greenwood and Wefelmeyer (1990a, relation (2.18) and Remark 2.19), an estimator $\hat{k}_n$ for k is regular and efficient at ϑ in the sense of a Hájek–Le Cam convolution theorem if and only if

$$c_n(\hat{k}_n - k(\vartheta)) = G_n(\vartheta) + o_{P_{n\vartheta}}(1) \qquad (1.4)$$

with

$$G_n(\vartheta) = c_n^{-1} \sum_{i=1}^n \int_0^1 D_{ni\,\vartheta}(g_\vartheta)(s)\, dM_{ni\,\vartheta}(s)$$

The gradient g_ϑ is the direction of steepest ascent of the functional k. Hence it is the least favorable direction for estimating k. We may think of $G_n(\vartheta)$ as the derivative of the log-likelihood in the least favorable direction. (Compare Remark 3.10.) In Section 4 we show that G_n admits the following expansion, for τ near ϑ,

$$G_n(\tau) = G_n(\vartheta) - c_n(k(\tau) - k(\vartheta)) + o_{P_{n\vartheta}}(1) \qquad (1.5)$$

This expansion follows from local asymptotic normality (1.2), and differentiability (1.3) of the functional. For fully specified semiparametric models based on i.i.d. observations, relation (1.5) is familiar in connection with Newton–Raphson improvement of estimators. See Bickel (1982, p. 648, Remark 1), Klaassen (1987, p. 1550, (2.2)) or Schick (1987, p. 91, (2.11)). We will see below that relation (1.5) is also useful for proving efficiency of nonparametric maximum likelihood estimators.

Uniform local asymptotic normality (1.2) and expansion (1.5) rely on a number of regularity conditions. Fortunately, we need not check them in specific examples. The main purpose of Sections 3 and 4 is to set out general conditions under which expansion (1.5) holds. In Sections 5 and 6 we simply take this expansion as an assumption. We will see that it is fairly easy to check in the examples of Sections 7 and 8.

We are now ready to describe three methods for constructing efficient estimators for a real-valued functional $k: \Theta \to \mathbb{R}$.

1.1 Martingale Estimators

Sometimes we can find estimators $\tilde{D}_{ni}$ for $D_{ni\,\vartheta}(g_\vartheta)$ such that

$$G_n(\vartheta) = c_n^{-1} \sum_{i=1}^{n} \int_0^1 \tilde{D}_{ni}(s)\,dM_{ni\,\vartheta}(s) + o_{P_{n\vartheta}}(1) \qquad (1.6)$$

and

$$c_n^{-2} \sum_{i=1}^{n} \int_0^1 \tilde{D}_{ni}(s)\,dM_{ni\,\vartheta}(s) = \hat{k}_n - \kappa(\vartheta) \qquad (1.7)$$

Equation (1.7) says that the difference between the estimator $\hat{k}_n$ and the value $k(\vartheta)$ of the functional can be written as a stochastic integral with respect to the martingale $M_{ni\,\vartheta}$. Such estimators are called *martingale estimators*. Relations (1.6) and (1.4) imply that $\hat{k}_n$ is efficient. Although the scope of this approach seems to be limited, it works in the examples of Sections 2 and 7.

Note that if (1.6) and (1.7) hold with ϑ replaced by τ, we obtain the expansion (1.5).

1.2 Newton–Raphson Improvement

Let $\tilde{k}_n$ be an estimator for k and $\tilde{G}_n$ an 'estimator' for G_n such that

$$\tilde{G}_n = G_n(\vartheta) - c_n(\tilde{k}_n - k(\vartheta)) + o_{P_{n\vartheta}}(1) \qquad (1.8)$$

Define an *improved estimator*

$$\hat{k}_n = \tilde{k}_n + c_n^{-1}\tilde{G}_n$$

Then $\hat{k}_n$ fulfills (1.4) and is, therefore, efficient.

This method is the most widely applicable of all three methods. One case is the following. Assume that the expansion (1.5) holds. Let $\tilde{\vartheta}_n$ be a c_n-consistent estimator for ϑ in the sense that $P_{n\vartheta} \circ c_n(\tilde{\vartheta}_n - \vartheta)$, $n \in \mathbb{N}$, is tight. Then (1.8) holds for $\tilde{k}_n = k(\tilde{\vartheta}_n)$ and $\tilde{G}_n = G_n(\tilde{\vartheta}_n)$.

1.3 Estimating Equations

Let $\hat{\vartheta}_n$ be a c_n-consistent solution of the estimating equation

$$G_n(\hat{\vartheta}_n) = o_{P_{n\vartheta}}(1) \qquad (1.9)$$

(It may be necessary to extend G_n from Θ to a larger set to get a simple solution.) Applying expansion (1.5) for τ replaced by $\hat{\vartheta}_n$, we obtain

(1.4), and hence efficiency, for the estimator $\hat{k}_n = k(\hat{\vartheta}_n)$.

Assume now that Θ is a *function space*, say a subset of the Skorohod space $D[0, 1]$. The methods described above can also be used to obtain efficient estimators for the parameter. We decompose the problem into a collection of problems, indexed by $t \in [0, 1]$. For fixed $t \in [0, 1]$, we consider the problem of estimating the real-valued functional

$$k_t(\vartheta) = \vartheta(t)$$

Each of the three methods gives an estimator $\hat{\vartheta}_n(t)$ of $\vartheta(t)$ fulfilling

$$c_n(\hat{\vartheta}_n(t) - \vartheta(t)) = G_{nt}(\vartheta) + o_{P_{n\vartheta}}(1)$$

with

$$G_{nt}(\vartheta) = c_n^{-1} \sum_{i=1}^{n} \int_0^1 D_{ni\,\vartheta}(g_{t\vartheta})(s)\, dM_{ni\,\vartheta}(s)$$

and with $g_{t\vartheta}$ the canonical gradient of k_t. Hence the estimator $\hat{\vartheta}_n$ for ϑ is efficient.

In this setting, the third method (based on estimating equations) proves efficiency of *nonparametric maximum likelihood estimators* in the sense discussed by Gill (1989, 1987). Let $\hat{\vartheta}_n$ be a c_n-consistent solution of the collection of estimating equations

$$G_{nt}(\hat{\vartheta}_n) = 0 \qquad t \in [0, 1] \tag{1.10}$$

Recall that $G_{nt}(\vartheta)$ can be thought of as (an extension of) the derivative of the log-likelihood in the direction $g_{t\vartheta}$ least favorable for the functional $k_t(\vartheta) = \vartheta(t)$. Hence we expect a nonparametric maximum likelihood estimator to fulfill (1.10). This implies efficiency. We compare our approach with Gill's in Section 2.

Note that in order to find a solution of (1.10), it is often necessary to extend the *derivatives* G_{nt} smoothly from Θ to a larger set. The common definition of nonparametric maximum likelihood estimators involves extending the likelihood function *itself* and maximizing this extension. As remarked by Gill (1989, p. 116), this may lead to inefficient and even inconsistent estimators if the extension of the likelihood does not entail a smooth extension of the derivative.

Assume that $\hat{\vartheta}_n$ is c_n-consistent and solves (1.10) only approximately:

$$G_{nt}(\hat{\vartheta}_n) = o_{P_{n\vartheta}}(1) \qquad t \in [0, 1] \tag{1.11}$$

This suffices for $\hat{\vartheta}_n$ to be efficient for ϑ. The converse is also true. Hence (1.11) is a *characterization* of regular and efficient estimators. To see this, let $\hat{\vartheta}_n$ be regular and efficient for ϑ. Then $\hat{\vartheta}_n(t)$ is regular and efficient for $\vartheta(t)$, for each $t \in [0, 1]$. Hence by (1.4),

$$c_n(\hat{\vartheta}_n(t) - \vartheta(t)) = G_{nt}(\vartheta) + o_{P_{n\vartheta}}(1)$$

Relation (1.11) now follows from expansion (1.5) for $k_t(\vartheta) = \vartheta(t)$.

Note that a solution of the *approximate* estimating equation (1.11) can usually be found *without* extending G_{nt} from Θ to a larger set. Extending G_{nt} and solving the exact equations (1.10) involves (at least) as much arbitrariness as solving the approximate estimating equations (1.11) without extending G_{nt}. For this reason, it may sometimes be convenient to avoid extending G_{nt} even though solutions of (1.11) cannot be interpreted as *exact* nonparametric maximum likelihood estimators.

The paper is organized as follows. In Section 2 we illustrate the three methods for constructing efficient estimators in the simple example of estimating the cumulative hazard rate on the basis of i.i.d. observations. In Section 3 we prove a uniform version of local asymptotic normality for the counting process model. In Section 4 we derive the expansion (1.5) for the derivative of the (partially specified) log-likelihood in the direction least favorable for a given functional. In Sections 5 and 6 we show efficiency of Newton–Raphson improvements and solutions of estimating equations, respectively.

We illustrate the three methods for constructing efficient estimators in counting process models in Sections 7 and 8. In Section 7 we consider Aalen's (1980) *additive risk model*, with intensity process

$$a_{ni\,\vartheta}(s, Y_{ni}(s)) = Y_{ni}(s)\lambda(s)$$

Here Y_{ni} is a row vector of predictable *covariate processes*, and λ is a (deterministic) vector of *hazard functions*. We are interested in efficient estimation of the *cumulative hazard rate*

$$\vartheta(t) = \int_0^t \lambda(s)\,ds$$

Aalen (1980) suggested a least squares estimator of ϑ. Huffer and McKeague (1990) determined its asymptotic covariance function and suggested a weighted least squares estimator with smaller covariance function. We obtain Huffer and McKeague's estimator as an efficient martingale estimator, and also interpret it as a nonparametric maximum likelihood estimator in the sense described above. We use Aalen's estimator as an initial estimator for the Newton–Raphson improvement procedure.

In Section 8 we consider Andersen and Gill's (1982) counting process version of Cox's (1972) *proportional hazards model*, with intensity process

$$a_{ni\,\vartheta}(s, Y_{ni}(s)) = C_{ni}(s)\lambda(s)\exp(\beta' Y_{ni}(s))$$

Here C_{ni} is a predictable *censoring process* taking only values 0 and 1, Y_{ni} is a vector of predictable *covariate processes*, β is a vector of *regression coefficients*, and λ is a (deterministic) *baseline hazard function*. We are interested in efficient estimation of the vector of regression coefficients. Andersen and Gill (1982) determined the asymptotic covariance of the estimator maximizing Cox's partial likelihood. We obtain Cox's estimator as an efficient martingale estimator, and also interpret it as a component of a nonparametric maximum likelihood estimator in the sense described above.

By convention, if a square matrix D is not invertible, set D^{-1} equal to the zero matrix.

2 A SIMPLE EXAMPLE

Consider the problem of estimating the *cumulative hazard rate*

$$\Lambda(t) = \int_0^t (1 - F(s))^{-1} f(s)\, ds$$

on the basis of n i.i.d. observations from an absolutely continuous distribution function F on the unit interval $[0, 1]$, with density f. This model can be written as a *multiplicative intensity model*: We observe a counting process $X_n = nF_n$ on $[0, 1]$, with F_n the empirical distribution function. Its intensity process is of the form $Y_n(s)\lambda(s)$ with $Y_n = n(1 - F_{n-})$ and $\lambda = (1 - F)^{-1}f$. The example is a special case of the multiplicative intensity model introduced by Aalen (1975, 1978). A multivariate generalization, the additive risk model of Aalen (1980), will be treated in Section 7.

Let the parameter space Θ be the set of cumulative hazard rates, Λ. We are interested in constructing efficient estimators for Λ.

In a first step, we prove that the model is locally asymptotically normal. Fix Λ. The log-likelihood between Λ' and Λ can be written as

$$\log(dP_{\Lambda'}^n/dP_\Lambda^n) = \int \log(\lambda'/\lambda)\, dX_n(s) - \int Y_n(\lambda' - \lambda)\, ds \quad (2.1)$$

Here and in the following, we omit the integration variable s. A *local model* may be introduced as follows. For any bounded function v on $[0, 1]$, define a sequence of parameters Λ_{nv} by

$$d\Lambda_{nv} = (1 + n^{-1/2}v)\, d\Lambda$$

By Taylor expansion of (2.1), for $\Lambda' = \Lambda_{nv}$, we obtain *local asymptotic normality*,

$$L_{nv} = \log(dP_{\Lambda_{nv}}^n/dP_\Lambda^n)$$

$$= n^{-1/2} \int v\, dM_{n\Lambda}(s) - \tfrac{1}{2}\|v\|_\Lambda^2 + o_{P_\Lambda}(1) \quad (2.2)$$

with

$$P_\Lambda^n \circ n^{-1/2} \int v\, dM_{n\Lambda}(s) \Rightarrow N(0, \|v\|_\Lambda^2)$$

Here $M_{n\Lambda}$ is the martingale defined by

$$dM_{n\Lambda}(s) = dX_n(s) - Y_n\lambda\, ds = n\, dF_n(s) - n(1 - F_{n-})\lambda\, ds$$

and the norm $\|\ \|_\Lambda$ is defined by

$$\|v\|_\Lambda^2 = \int v^2 f\, ds$$

We call the corresponding inner product $(\ ,\ )_\Lambda$ the *acuity*.

In a second step, we decompose the problem of estimating the parameter Λ into a collection of problems, indexed by $t \in [0, 1]$. For fixed $t \in [0, 1]$, we consider the problem of estimating the real-valued functional

$$k_t(\Lambda) = \Lambda(t) = \int_0^t \lambda\, ds$$

The functional k_t is linear in the local parameter v and can be expressed in terms of a *gradient*, $g_{t\Lambda}$, with respect to the acuity:

$$n^{1/2}(k_t(\Lambda_{n\nu}) - k_t(\Lambda)) = n^{1/2}(\Lambda_{n\nu}(t) - \Lambda(t))$$

$$= \int_0^t \nu\lambda \, ds = (\nu, g_{t\Lambda})_\Lambda$$

with

$$g_{t\Lambda}(s) = (1 - F(s))^{-1}1_{[0,t]}(s) \qquad s \in [0, 1]$$

The gradient $g_{t\Lambda}$ is the direction of steepest ascent of k_t. Hence it is the least favorable direction for estimating k_t.

According to Greenwood and Wefelmeyer (1990a, relation (2.18) and Remark 2.19), an estimator $\hat{\Lambda}_n(t)$ of $\Lambda(t)$ is regular and efficient in the sense of a Hájek–Le Cam convolution theorem if and only if

$$n^{1/2}(\hat{\Lambda}_n(t) - \Lambda(t)) = G_{nt}(\Lambda) + o_{P_\Lambda}(1) \qquad (2.3)$$

with

$$G_{nt}(\Lambda) = n^{-1/2} \int g_t \, dM_{n\Lambda}(s)$$

$$= n^{1/2} \int_0^t (1 - F)^{-1}(dF_n(s) - (1 - F_{n-})\lambda \, ds) \qquad (2.4)$$

Replace F by a (predictable) estimator F_{n-} to obtain

$$G_{nt}(\Lambda) = \tilde{G}_{nt}(\Lambda(t)) + o_{P_\Lambda}(1) \qquad (2.5)$$

with

$$\tilde{G}_{nt}(k) = n^{1/2}\left(\int_0^t (1 - F_{n-})^{-1} \, dF_n(s) - k\right) \qquad k \in \mathbb{R} \qquad (2.6)$$

2.1 Martingale Estimator

Comparing (2.5) with (2.3), we see that an efficient estimator $\hat{\Lambda}_n(t)$ for $\Lambda(t)$ is given by

$$\hat{\Lambda}_n(t) = \int_0^t (1 - F_{n-})^{-1} \, dF_n(s) \qquad (2.7)$$

This is the *Nelson estimator*. Introduce

$$\Lambda_n^*(t) = \int_0^t 1\{F_{n-} < 1\}\lambda \, ds$$

This is a slight modification of $\Lambda(t)$. The difference between $\hat{\Lambda}_n(t)$ and $\Lambda_n^*(t)$ can be written as a stochastic integral with respect to $M_{n\Lambda}$,

$$\hat{\Lambda}_n(t) - \Lambda_n^*(t) = n^{-1} \int_0^t (1 - F_{n-})^{-1} \, dM_{n\Lambda}(s)$$

In this sense, $\hat{\Lambda}_n(t)$ is a *martingale estimator*. (Recall that, by convention, $(1 - F_{n-})^{-1} = 0$ if $F_{n-} = 1$.)

Let us also illustrate the two other approaches described in the Introduction.

2.2 Newton–Raphson Improvement

If we have a preliminary estimator $\tilde{\Lambda}_n(t)$ for $\Lambda(t)$, we can use a version of the Newton–Raphson improvement. Estimate $G_{nt}(\Lambda)$ by $\tilde{G}_{nt}(\tilde{\Lambda}_n(t))$, with $\tilde{G}_{nt}$ defined in (2.6). From (2.5),

$$\tilde{G}_{nt}(\tilde{\Lambda}_n(t)) = \tilde{G}_{nt}(\Lambda(t)) - n^{1/2}(\tilde{\Lambda}_n(t) - \Lambda(t))$$
$$= G_{nt}(\Lambda) - n^{1/2}(\tilde{\Lambda}_n(t) - \Lambda(t)) + o_{P_\Lambda}(1)$$

Hence (2.3) is fulfilled for the improved estimator

$$\tilde{\tilde{\Lambda}}_n(t) = \tilde{\Lambda}_n(t) - n^{-1/2}\tilde{G}_{nt}(\tilde{\Lambda}_n(t)) \tag{2.8}$$

This implies that the improved estimator is efficient. Note that, by (2.6), the improved estimator reproduces the Nelson estimator (2.7) *exactly*.

2.3 Estimating Equation

The Nelson estimator (2.7) can also be interpreted as nonparametric maximum likelihood estimator. Extend k_t and G_{nt} to $D[0, 1]$ as follows. The functional $k_t(\Lambda) = \Lambda(t)$ has an obvious extension, evaluation at t. For $M \notin \Theta$ set

$$G_{nt}(M) = \tilde{G}_{nt}(M(t)) \tag{2.9}$$

The Nelson estimator is a pure jump process. Hence it is outside Θ with $P_{n\vartheta}$-probability tending to 1. From definition (2.6) of $\tilde{G}_{nt}$ we see that the Nelson estimator solves, simultaneously for each $t \in [0, 1]$, the estimating equation

$$G_{nt}(\hat{\Lambda}_n) = 0 \tag{2.10}$$

As noted in Section 1, we can think of $G_{nt}(\Lambda)$ as (an extension of) the derivative of the log-likelihood in the direction least favorable for the functional $k_t(\Lambda) = \Lambda(t)$. Hence an estimator solving (2.10) for all t may be interpreted as a maximum likelihood estimator.

Relation (2.3), and hence efficiency of $\hat{\Lambda}_n(t)$, now follows since by (2.5) and (2.6) we have for $M \notin \Theta$,

$$
\begin{aligned}
G_{nt}(M) - G_{nt}(\Lambda) &= \tilde{G}_{nt}(M(t)) - G_{nt}(\Lambda) \\
&= \tilde{G}_{nt}(M(t)) - \tilde{G}_{nt}(\Lambda(t)) + o_{P_\Lambda}(1) \\
&= -n^{1/2}(M(t) - \Lambda(t)) + o_{P_\Lambda}(1)
\end{aligned}
$$

(This is expansion (1.5).)

Our argument for efficiency of nonparametric maximum likelihood estimators is related to that of Gill (1989, 1987). The main difference is that Gill does not reduce the determination of the asymptotic distribution of $\hat{\Lambda}_n$ to a collection of one-dimensional problems, consisting of determining the asymptotic distribution of $\hat{\Lambda}_n(t)$ for each t separately. Let us describe Gill's approach for our example. The space of all possible directions at Λ in the parameter space is spanned by the indicator functions $1_{[0,t]}$, $t \in [0, 1]$. The derivative of the log-likelihood in direction $1_{[0,t]}$ is (see (2.2))

$$
H_{nt}(\Lambda) = n^{-1/2} M_{n\Lambda}(t) = n^{1/2} \int_0^t (dF_n(s) - (1 - F_{n-}) \, d\Lambda(s))
$$

The random function H_{nt} has a natural extension from Θ to $D[0, 1]$. The Nelson estimator (2.7) solves, simultaneously for each $t \in [0, 1]$, the estimating equation

$$
H_{nt}(\hat{\Lambda}_n) = 0
$$

Consider the operator, H_n, on $D[0, 1]$ which maps Λ into the function $t \rightarrow H_{nt}(\Lambda)$. Then $H_n(\hat{\Lambda}_n) = 0$, and one can try to obtain the asymptotic distribution of $\hat{\Lambda}_n$ by a Taylor expansion around Λ, just as in the classical proof for finite-dimensional parameters. As Gill observes, there are, however, technical problems in making the heuristic reasoning rigorous. In particular, the sample Fisher information matrix is now replaced by a random linear operator which is difficult to invert.

Let us point out a restriction in Gill's approach which is, perhaps, not quite obvious. The class of directions $1_{[0,t]}$, $t \in [0, 1]$, at Λ is chosen independently of the underlying parameter Λ. This is possible because of a special feature of the parameter space in the example. (This feature is, in fact, also present in the examples of Sections 7 and 8.) Such a choice is impossible in many other (infinite-dimensional) examples. Consider, for example, the problem of estimating the distribution function F. Fix F, and define a sequence F_{nv} of distribution functions by

$$dF_{nv} = (1 + n^{-1/2}v)\,dF$$

Integrating on both sides, we see that the direction v has expectation 0 under F. Hence no v can be a direction at *every* distribution function F. The point here is that the set of all distribution functions is a subset of a sphere and therefore curved. In infinite-dimensional models, curved parameter spaces are common.

3 LOCAL ASYMPTOTIC NORMALITY

Let D be a topological vector space and $\Theta \subset D$ a parameter space. Fix $n \in \mathbb{N}$. Consider a model

$$(\Omega_n, \mathscr{F}_n, P_{n\vartheta} \colon \vartheta \in \Theta)$$

Suppose we observe counting processes

$$X_{n1}(t), \ldots, X_{nn}(t) \qquad t \in [0, 1]$$

Assume that the X_{ni} are $P_{n\vartheta}$-a.s. finite, and that no two of them jump simultaneously. Set $X_n = (X_{n1}, \ldots, X_{nn})$. Let $\mathbb{F}_n = (\mathscr{F}_{nt})_{t\in[0,1]}$ denote the filtration generated by X_n. Let $\mathbb{G}_n = (\mathscr{G}_{nt})_{t\in[0,1]}$ be a possibly larger filtration. Assume that the Doob–Meyer decomposition of X_{ni} with respect to $\mathbb{G}_n$ and $P_{n\vartheta}$ is of the form

$$X_{ni}(t) = M_{ni\vartheta}(t) + \int_0^t a_{ni\vartheta}(s)\,ds \qquad t \in [0, 1] \qquad (3.1)$$

with $M_{ni\vartheta}$ a martingale and $a_{ni\vartheta}$ a predictable *intensity process*. In applications, $a_{ni\vartheta}$ will be a deterministic function of a predictable covariate process Y_{ni} on $[0, 1]$. See Sections 7 and 8.

We introduce a *partially specified log-likelihood* between δ and τ,

$$L_{n\tau\delta} = \sum_{i=1}^n \left(\int_0^1 \log(a_{ni\delta}/a_{ni\tau})\,dX_{ni}(s) - \int_0^1 (a_{ni\delta} - a_{ni\tau})\,ds \right) \quad (3.2)$$

If $\mathbb{G}_n = \mathbb{F}_n$, this is the usual log-likelihood.

Let A_n be a sequence of subsets of Θ and V_n a sequence of sets. We say that

$$f_{n\tau v} \to 0 \qquad (P_{n\tau}) \qquad \text{uniformly for } \tau \in A_n \text{ and } v \in V_n$$

if, for each $\epsilon > 0$,

$$\sup_{\tau \in A_n} P_{n\tau} \{ \sup_{v \in V_n} |f_{n\tau v}| > \epsilon \} \to 0$$

We say that

$$Q_{n\tau v} - R_{n\tau v} \Rightarrow 0 \qquad \text{uniformly for } \tau \in A_n \text{ and } v \in V_n$$

if, for each bounded continuous function f,

$$\sup_{\tau \in A_n, v \in V_n} \int f(dQ_{n\tau v} - dR_{n\tau v}) \to 0$$

The following conditions imply a uniform and infinite-dimensional version of local asymptotic normality for the (partially specified) likelihood. Let H_τ be a Hilbert space with inner product $(\, , \,)_\tau$ and norm $\| \; \|_\tau$. Introduce a *local parameterization*

$$\xi_{n\tau} : \Theta \to H_\tau$$

We assume that $\xi_{n\tau}(\delta)$ stays in a $\| \; \|_\tau$-bounded set uniformly for $c_n(\tau - \vartheta)$ and $c_n(\delta - \vartheta)$ in compact sets. The map $\xi_{n\tau}$ will rescale $\delta \in \Theta$ around τ as $c_n(\delta - \tau)$ with $c_n \to \infty$ and then apply an approximate linear isometry from D into H_t.

For $i = 1, \ldots, n$ let

$$D_{ni\tau} : H_\tau \times \Omega_n \times [0, 1] \to \mathbb{R}$$

be linear in the first variable, and adapted to $\mathbb{G}_n$ for fixed first variable. We assume the following conditions.

1. The intensity processes $a_{ni\delta}$ are *Hellinger differentiable* at $\delta = \tau$ with *differential operator* $D_{ni\tau}$: Uniformly for $c_n(\tau - \vartheta)$ and $c_n(\delta - \vartheta)$ in compact sets,

$$\sum_{i=1}^{n} \int_0^1 ((a_{ni\delta}/a_{ni\tau})^{1/2} - 1 - \tfrac{1}{2}c_n^{-1} D_{ni\tau}(\xi_{n\tau}(\delta)))^2 a_{ni\tau} \, ds$$

$$\to 0 \qquad (P_{n\tau}) \tag{3.3}$$

2. The differential operators $D_{ni\tau}$ fulfill a *Lindeberg condition*: Uniformly for $c_n(\tau - \vartheta)$ in compact sets and $\|v\|_\tau$ bounded,

$$c_n^{-2} \sum_{i=1}^{n} \int_0^1 D_{ni\tau}(v)^2 1\{|D_{ni\tau}(v)| > \epsilon c_n\} a_{ni\tau} \, ds \to 0 \qquad (P_{n\tau}) \; (3.4)$$

3. The Hellinger sequence is *asymptotically nonrandom*: uniformly for $c_n(\tau - \vartheta)$ in compact sets and $\|v\|_\tau$ bounded,

$$c_n^{-2} \sum_{i=1}^n \int_0^1 D_{ni\tau}(v)^2 a_{ni\tau}\,ds - \|v\|_\tau^2 \to 0 \qquad (P_{n\tau}) \qquad (3.5)$$

It is asymptotic nonrandomness (3.5) which makes the model locally asymptotically normal as opposed to locally asymptotically mixed normal. We impose the condition for simplicity. For independent and identically distributed processes, conditions (3.3) to (3.5) simplify; see Greenwood and Wefelmeyer (1990b, Section 2). The rate c_n is then the familiar rate $n^{1/2}$.

(3.6) Proposition. If (3.3) to (3.5) hold, then, uniformly for $c_n(\tau - \vartheta)$ and $c_n(\delta - \vartheta)$ in compact sets,

$$L_{n\tau\delta} - Z_{n\tau}(\xi_{n\tau}(\delta)) + \tfrac{1}{2}\|\xi_{n\tau}(\delta)\|_\tau^2 \to 0 \qquad (P_{n\tau}) \qquad (3.7)$$

with

$$Z_{n\tau}(v) = c_n^{-1} \sum_{i=1}^n \int_0^1 D_{ni\tau}(v)\,dM_{ni\tau}(s)$$

Furthermore, uniformly for $c_n(\tau - \vartheta)$ in compact sets and $\|v\|_\tau$ bounded,

$$P_{n\tau} \circ Z_{n\tau}(v) - N(0, \|v\|_\tau^2) \Rightarrow 0 \qquad (3.9)$$

Proof. To prove (3.7), replace the log in the (partially specified) log-likelihood (3.2) by

$$\log x = 2\log(1 + x^{1/2} - 1)$$
$$= 2(x^{1/2} - 1) - (x^{1/2} - 1)^2 + 2r(x^{1/2} - 1)$$

and use $x - 1 = 2(x^{1/2} - 1) + (x^{1/2} - 1)^2$ to obtain

$$L_{n\tau\delta} = 2\sum \int ((a_{ni\delta}/a_{ni\tau})^{1/2} - 1)\,dM_{ni\tau}(s)$$
$$- 2\sum \int ((a_{ni\delta}/a_{ni\tau})^{1/2} - 1)a_{ni\tau}\,ds + R_{n\tau\delta}$$

For the remainder we have $R_{n\tau\delta} \to 0$ $(P_{n\tau})$ uniformly for $c_n(\tau - \vartheta)$ and $c_n(\delta - \vartheta)$ in compact sets by noticing that the arguments in Greenwood and Wefelmeyer (1990a, Section 2) apply uniformly. Next replace the expressions $(a_{ni\,\delta}/a_{ni\,\tau})^{1/2} - 1$ in the first sum by $\frac{1}{2}c_n^{-1}D_{ni\,\tau}(\xi_{n\tau}(\delta))$, using (3.3). Replace the second sum by $\frac{1}{4}\|\xi_{n\tau}(\delta)\|_\tau^2$ using (3.3) and (3.5) with v replaced by $\xi_{n\tau}(\delta)$.

Relation (3.9) follows from a uniform version of the central limit theorem for semimartingales; see Greenwood and Shiryaev (1990).

(3.10) Remark. Heuristically, $Z_{n\tau}$ is the differential operator of the partially specified log-likelihood, considered as a function of the local parameter $\xi_{n\tau}(\delta)$. This is true (in an asymptotic and stochastic sense) if (3.7) holds with $o_{P_{n\tau}}(1)$ replaced by the stronger $o_{P_{n\tau}}(\|\xi_{n\tau}(\delta)\|_\tau)$. For *fully* specified general semimartingale models with *finite*-dimensional parameter, Jacod (1989, p. 463, Theorem 6.2) gives necessary and sufficient conditions for *nonasymptotic* Hellinger differentiability of likelihoods. In Sections 7 and 8 we will base construction of efficient estimators on $Z_{n\tau}$. The arguments work because $Z_{n\tau}$ appears as linear process in the stochastic expansion (3.7) of the likelihood. We do not need that $Z_{n\tau}$ is a differential operator of the likelihood. This is, however, helpful if we want to interpret the estimators as nonparametric maximum likelihood estimators.

4 QUADRATIC APPROXIMATIONS

In this section we set out conditions under which the basic expansion (1.5) holds. Fix $\vartheta \in \Theta$. In the following, relations involving τ or δ are meant *uniformly* for $c_n(\tau - \vartheta)$ or $c_n(\delta - \vartheta)$ in compact sets.

Assume that the local asymptotic normality assertions of Proposition 3.6 hold. If the model were fully specified, we would have contiguity between $P_{n\tau}$ and $P_{n\vartheta}$ in this range. Since the model is only partially specified, contiguity must be assumed.

Choose a vector field $f_\tau \in H_\tau$ for $\tau \in \Theta$. Assume that the vector field is *tangent* to Θ in the sense that for $\tau \in \Theta$ there exist $\delta_{n\tau} \in \Theta$, with $c_n(\delta_{n\tau} - \vartheta)$ staying in a compact set, such that

$$\|f_\tau - \xi_{n\tau}(\delta_{n\tau})\|_\tau \to 0 \tag{4.1}$$

Assume that the vector field depends smoothly on the parameter τ in the sense that

$$\|\xi_{n\vartheta}(\delta_{n\tau}) - \xi_{n\vartheta}(\tau) - f_\vartheta\|_\vartheta \to 0 \tag{4.2}$$

and

$$\|f_\tau\|_\tau \to \|f_\vartheta\|_\vartheta \qquad (4.3)$$

From (4.2) and (3.9) with $\tau = \vartheta$ and $v = \xi_{n\vartheta}(\delta_{n\tau}) - \xi_{n\vartheta}(\tau) - f_\vartheta$ we have

$$Z_{n\vartheta}(\xi_{n\vartheta}(\delta_{n\tau})) = Z_{n\vartheta}(\xi_{n\vartheta}(\tau) + f_\vartheta) + o_{P_{n\vartheta}}(1) \qquad (4.4)$$

By definition (3.2),

$$L_{n\tau\delta_{n\tau}} = L_{n\vartheta\delta_{n\tau}} - L_{n\vartheta\tau}$$

Relations (4.2) and (3.9) imply

$$Z_{n\vartheta}(\xi_{n\vartheta}(\delta_{n\tau})) - Z_{n\vartheta}(\xi_{n\vartheta}(\tau)) = Z_{n\vartheta}(f_\vartheta) + o_{P_{n\vartheta}}(1)$$

From (4.2), again, we now obtain

$$\begin{aligned}
L_{n\tau\delta_{n\tau}} &= Z_{n\vartheta}(\xi_{n\vartheta}(\delta_{n\tau})) - \tfrac{1}{2}\|\xi_{n\vartheta}(\delta_{n\tau})\|_\vartheta^2 \\
&\quad - Z_{n\vartheta}(\xi_{n\vartheta}(\tau)) + \tfrac{1}{2}\|\xi_{n\vartheta}(\tau)\|_\vartheta^2 + o_{P_{n\vartheta}}(1) \\
&= Z_{n\vartheta}(f_\vartheta) - \tfrac{1}{2}\|\xi_{n\vartheta}(\tau) + f_\vartheta\|_\vartheta^2 \\
&\quad + \tfrac{1}{2}\|\xi_{n\vartheta}(\tau)\|_\vartheta^2 + o_{P_{n\vartheta}}(1) \\
&= Z_{n\vartheta}(f_\vartheta) - (\xi_{n\vartheta}(\tau), f_\vartheta)_\vartheta - \tfrac{1}{2}\|f_\vartheta\|_\vartheta^2 + o_{P_{n\vartheta}}(1) \quad (4.5)
\end{aligned}$$

This is an infinite-dimensional version of a quadratic approximation in the sense of Le Cam (1986, p. 210, Definition 1), with $f \to f'X$ replaced by

$$f \to Z_{n\vartheta}(f) - (\xi_{n\vartheta}(\tau), f)$$

(For finite-dimensional models, Basawa and Koul (1988) use quadratic approximations to derive asymptotic distributions for several types of estimators.)

Because of (4.1), (3.9) and contiguity,

$$Z_{n\tau}(\xi_{n\tau}(\delta_{n\tau})) = Z_{n\tau}(f_\tau) + o_{P_{n\vartheta}}(1) \qquad (4.6)$$

Hence local asymptotic normality (3.7) for $\delta = \delta_{n\tau}$, contiguity and relation (4.1) imply

$$\begin{aligned}
L_{n\tau\delta_{n\tau}} &= Z_{n\tau}(\xi_{n\tau}(\delta_{n\tau})) - \tfrac{1}{2}\|\xi_{n\tau}(\delta_{n\tau})\|_\tau^2 + o_{P_{n\vartheta}}(1) \\
&= Z_{n\tau}(f_\tau) - \tfrac{1}{2}\|f_\vartheta\|_\vartheta^2 + o_{P_{n\vartheta}}(1) \qquad (4.7)
\end{aligned}$$

Comparing (4.5) and (4.7), we have

$$Z_{n\tau}(f_\tau) = Z_{n\vartheta}(f_\vartheta) - (\xi_{n\vartheta}(\tau), f_\vartheta)_\vartheta + o_{P_{n\vartheta}}(1) \qquad (4.8)$$

Note that the parameter τ enters the right-hand approximation only through the local parameterization $\xi_{n\vartheta}(\tau)$, a *deterministic* function of τ.

Consider now a real-valued functional

$$k: \Theta \to \mathbb{R}$$

Assume that the functional is *differentiable* at ϑ with *gradient* $g_\vartheta \in H_\vartheta$ in the sense that uniformly for $c_n(\tau - \vartheta)$ in compact sets,

$$c_n(k(\tau) - k(\vartheta)) - (\xi_{n\vartheta}(\tau), g_\vartheta)_\vartheta \to 0 \qquad (4.9)$$

Comparing (4.8) and (4.9), we see that

$$Z_{n\tau}(f_\tau) = Z_{n\vartheta}(f_\vartheta) - c_n(k(\tau) - k(\vartheta)) + o_{P_{n\vartheta}}(1) \qquad (4.10)$$

if and only if

$$(\xi_{n\vartheta}(\tau), f_\vartheta - g_\vartheta)_\vartheta \to 0 \qquad (4.11)$$

For $f_\vartheta = g_\vartheta$, relation (4.10) is the basic expansion (1.5). Versions of relation (4.10) will be used as assumptions in Sections 5 and 6. The purpose of Sections 3 and 4 was to motivate these assumptions.

5   IMPROVEMENT OF ESTIMATORS

The results of Sections 3 and 4 motivate the following simple versions of the Newton–Raphson improvement procedure for estimators. It is convenient to phrase the results in an abstract setting.

Let D be a topological vector space, $\Theta \subset D$ a parameter space, and $k: \Theta \to \mathbb{R}$ a functional. Consider a sequence of models

$$(\Omega_n, \mathcal{F}_n, P_{n\vartheta}: \vartheta \in \Theta) \qquad n \in \mathbb{N}$$

Fix $\vartheta \in \Theta$. For $\tau \in \Theta$ let

$$G_n(\tau): \Omega_n \to \mathbb{R}$$

A naive approach to improvement of estimators can be based on the following expansion, which is motivated by relation (4.10). Assume that uniformly for $\tau \in \Theta$ with $c_n(\tau - \vartheta)$ in compact sets,

$$G_n(\tau) = G_n(\vartheta) - c_n(k(\tau) - k(\vartheta)) + o_{P_{n\vartheta}}(1) \qquad (5.1)$$

Let $\tilde{\vartheta}_n: \Omega_n \to \Theta$, $n \in \mathbb{N}$, be an estimator-sequence which is c_n-*consistent* for ϑ in the sense that

$$P_{n\vartheta} \circ c_n(\tilde{\vartheta}_n - \vartheta) \quad \text{is tight} \tag{5.2}$$

Define an *improved estimator*

$$\hat{k}_n = k(\tilde{\vartheta}_n) + c_n^{-1} G_n(\tilde{\vartheta}_n)$$

Relations (5.1) and (5.2) now imply

$$c_n(\hat{k}_n - k(\vartheta)) = c_n(k(\tilde{\vartheta}_n) - k(\vartheta)) + G_n(\tilde{\vartheta}_n)$$
$$= G_n(\vartheta) + o_{P_{n\vartheta}}(1)$$

This approach requires relation (5.1) and a c_n-consistent estimator for ϑ. A more generally applicable method requires only a (c_n-consistent) estimator for k, together with an 'estimator' for G_n.

Let $\tilde{k}_n: \Omega_n \to \mathbb{R}$ be an estimator for k and $\tilde{G}_n$ an 'estimator' for G_n such that

$$\tilde{G}_n = G_n(\vartheta) - c_n(\tilde{k}_n - k(\vartheta)) + o_{P_{n\vartheta}}(1) \tag{5.3}$$

Define an *improved estimator*

$$\hat{k}_n = \tilde{k}_n + c_n^{-1} \tilde{G}_n$$

Then

$$c_n(\hat{k}_n - k(\vartheta)) = c_n(\tilde{k}_n - k(\vartheta)) + \tilde{G}_n$$
$$= G_n(\vartheta) + o_{P_{n\vartheta}}(1) \tag{5.4}$$

(5.5) Remark. In Sections 7 and 8 we will apply the improvement procedure for the counting process setting of Section 3. Consider a (partially specified) likelihood which is locally asymptotically normal with linear process $Z_{n\tau}$ at $\tau \in \Theta$. Let $k: \Theta \to \mathbb{R}$ be a functional which is differentiable, with gradient $g_\tau \in V_\tau$, at $\tau \in \Theta$. Set

$$G_n(\tau) = Z_{n\tau}(g_\tau)$$

Fix $\vartheta \in \Theta$. Find estimators $\tilde{k}_n$ for k and $\tilde{G}_n$ for G_n such that

$$\tilde{G}_n = Z_{n\vartheta}(g_\vartheta) - c_n(\tilde{k}_n - k(\vartheta)) + o_{P_{n\vartheta}}(1) \tag{5.6}$$

Define the *improved estimator*

$$\hat{k}_n = \tilde{k}_n + c_n^{-1} \tilde{G}_n$$

Relation (5.4) then implies

$$c_n(\hat{k}_n - k(\vartheta)) = Z_{n\vartheta}(g_\vartheta) + o_{P_{n\vartheta}}(1) \tag{5.7}$$

According to Greenwood and Wefelmeyer (1990a, relation (2.18) and Remark 2.19), relation (5.7) is necessary and sufficient for $\hat{k}_n$ to be a regular and efficient estimator for the functional k at ϑ.

6 EFFICIENT ESTIMATING EQUATIONS

The results of Sections 3 and 4 motivate the following simple proposition on solutions of estimating equations. As in Section 5, we phrase the result in an abstract setting.

Let D be topological vector space, $k: D \to \mathbb{R}$ a functional, $\Theta \subset D$ a parameter space. Consider a sequence of models

$$(\Omega_n, \mathscr{F}_n, P_{n\vartheta}: \vartheta \in \Theta) \qquad n \in \mathbb{N}$$

Fix $\vartheta \in \Theta$. For $\tau \in D$ let

$$G_n(\tau): \Omega_n \to \mathbb{R}$$

be such that uniformly for $c_n(\tau - \vartheta)$ in compact sets,

$$G_n(\tau) = G_n(\vartheta) - c_n(k(\tau) - k(\vartheta)) + o_{P_{n\vartheta}}(1) \qquad (6.1)$$

This relation is motivated by (4.10)

Let $\hat{\vartheta}_n: \Omega_n \to D$, $n \in \mathbb{N}$, be an estimator-sequence which is c_n-consistent for ϑ in the sense that

$$P_{n\vartheta} \circ c_n(\hat{\vartheta}_n - \vartheta) \quad \text{is tight} \qquad (6.2)$$

Assume that $\hat{\vartheta}_n$ is an approximate solution of the estimating equation

$$G_n(\hat{\vartheta}_n) = o_{P_{n\vartheta}}(1) \qquad (6.3)$$

There follows a stochastic expansion of the estimator $k(\hat{\vartheta}_n)$.

(6.4) Proposition. Assume (6.1) to (6.3). Then

$$c_n(k(\hat{\vartheta}_n) - k(\vartheta)) = G_n(\vartheta) + o_{P_{n\vartheta}}(1) \qquad (6.5)$$

Proof. Let $\epsilon > 0$. By tightness (6.2), there exists a compact set K such that for $n \in \mathbb{N}$,

$$P_{n\vartheta}\{c_n(\hat{\vartheta}_n - \vartheta) \in K\} \geq 1 - \epsilon$$

Together with (6.1) and (6.3) this implies for n sufficiently large,

$$P_{n\vartheta}\{|G_n(\vartheta) - c_n(k(\hat{\vartheta}_n) - k(\vartheta))| \geq \epsilon\} \leq 2\epsilon$$

Hence the assertion.

(6.6) Remark. The solution of the estimating equation (6.3) often falls outside the parameter space Θ, and even

$$P_{n\vartheta}\{\hat{\vartheta}_n \in \Theta\} \to 0$$

Then for (6.5) to be true it suffices that the expansion (6.1) holds only for $\tau \notin \Theta$. This is the case for the examples in Sections 7 and 8.

(6.7) Remark. In Sections 7 and 8 we will apply Proposition 6.4 for the counting process setting of Section 3. Consider a (partially specified) likelihood which is locally asymptotically normal with linear process $Z_{n\tau}$ at $\tau \in \Theta$. Let $k: \Theta \to \mathbb{R}$ be a functional which is differentiable, with gradient $g_\tau \in V_\tau$, at $\tau \in \Theta$. Set

$$G_n(\tau) = Z_{n\tau}(g_\tau).$$

Extend G_n and k to D. Fix $\vartheta \in \Theta$. Check (6.1). Let $\hat{\vartheta}_n: \Omega_n \to D, n \in \mathbb{N}$, be a c_n-consistent estimator-sequence for ϑ which solves the estimating equation

$$G_n(\hat{\vartheta}_n) = o_{P_{n\vartheta}}(1)$$

Proposition 6.4 then implies

$$c_n(k(\hat{\vartheta}_n) - k(\vartheta)) = Z_{n\vartheta}(g_\vartheta) + o_{P_{n\vartheta}}(1) \tag{6.8}$$

According to Greenwood and Wefelmeyer (1990a, Section 2), relation (6.8) is necessary and sufficient for $k(\hat{\vartheta}_n)$ to be a regular and efficient estimator for the functional k at ϑ.

7 AALEN'S ADDITIVE RISK MODEL

The *additive risk model* introduced by Aalen (1980) is of the form considered in Section 3, with intensity process

$$a_{ni\,\vartheta}(s) = Y_{ni}(s)\lambda(s) \qquad s \in [0, 1] \tag{7.1}$$

Here Y_{ni} is a p-dimensional row vector of predictable *covariate processes*, and λ is a p-dimensional column vector of bounded *hazard functions*.

Parametrize by the vector of *cumulative hazard rates* ϑ, defined by

$$\vartheta(t) = \int_0^t \lambda \, ds$$

Fix ϑ. Introduce *weights*

$$W_{ni\,\vartheta} = (Y_{ni}\lambda)^{-1}$$

Define

$$`W_{n\vartheta} = \text{diag } W_{ni\vartheta} \qquad \Lambda_\vartheta = \text{diag } \lambda_j$$

Let Y_n denote the matrix with rows Y_{ni}. Assume that there exists a matrix function U_ϑ with largest and smallest eigenvalue bounded and bounded away from 0, respectively, such that

$$\sup_{t \in [0,1]} |c_n^{-2} Y_n' W_{n\vartheta} Y_n - U_\vartheta| \to 0 \qquad (P_{n\vartheta}) \qquad (7.2)$$

We will apply the arguments of Sections 5 and 6 to construct efficient estimators for the cumulative hazard rate ϑ. For this purpose, we will need to verify the expansions (5.1) and (6.1). This will involve the linear process $Z_{n\vartheta}$ in the stochastic approximation (3.7) of the (partially specified) log-likelihood (3.2). The form of $Z_{n\vartheta}$ can be obtained from a nonuniform version of local asymptotic normality, as in Greenwood and Wefelmeyer (1990a, Section 3).

We introduce a local model at ϑ as follows. Let V be the set of all p-dimensional vectors of bounded functions on $[0, 1]$. For $v = (v_1, \ldots, v_p)' \in V$ and n sufficiently large define the sequence $\vartheta_{nv} = (\vartheta_{nv1}, \ldots, \vartheta_{nvp})'$ by

$$d\vartheta_{nvj} = (1 + c_n^{-1} v_j) \, d\vartheta_j$$

According to Greenwood and Wefelmeyer (1990a, Section 3), the (partially specified) log-likelihood (3.2) is *locally asymptotically normal* in the following sense:

$$L_{n\vartheta\vartheta_{nv}} = Z_{n\vartheta}(v) - \tfrac{1}{2}\|v\|_\vartheta^2 + o_{P_{n\vartheta}}(1)$$

Here the linear process $Z_{n\vartheta}$ is

$$Z_{n\vartheta}(v) = c_n^{-1} \int v' \Lambda_\vartheta Y_n' W_{n\vartheta} \, dM_{n\vartheta}(s)$$

with the martingale $M_{n\vartheta}$ defined by

$$dM_{n\vartheta}(s) = dX_n(s) - Y_n \, d\vartheta(s)$$

The norm $\| \; \|_\vartheta$ is defined by

$$\|v\|_\vartheta^2 = \int v' \Lambda_\vartheta U_\vartheta v \, ds$$

The *acuity* is defined as the inner product corresponding to this norm,

$$(v, w)_\vartheta = \int v' \Lambda_\vartheta U_\vartheta \Lambda_\vartheta w \, ds \qquad (7.3)$$

We are interested in efficient estimation of ϑ. We reduce the problem to estimating, for each $t \in [0, 1]$, the p-dimensional functional

$$k_t(\vartheta) = \vartheta(t)'$$

It is convenient to write the functional as a *row* vector. We have

$$c_n(k_t(\vartheta_{nv}) - k_t(\vartheta)) = c_n(\vartheta_{nv}(t) - \vartheta(t))' = \int_0^t v' \Lambda_\vartheta \, ds$$

Following Greenwood and Wefelmeyer (1990a, Section 3) we express the linear functional

$$v \to \int_0^t v' \Lambda_\vartheta \, ds$$

on the local parameter space V in terms of the acuity (7.3):

$$\int_0^t v' \Lambda_\vartheta \, ds = (v, g_{t\vartheta})_\vartheta \qquad \text{for } v \in V$$

with *gradient*

$$g_{t\vartheta}(s) = \Lambda_\vartheta(s)^{-1} U_\vartheta(s)^{-1} 1_{[0,t]}(s) \qquad s \in [0, 1]$$

As noted in Remark 5.5, a regular and efficient estimator $\hat{\vartheta}_n(t)$ for $\vartheta(t)$ is characterized by

$$c_n(\hat{\vartheta}_n(t) - \vartheta(t)) = G_{nt}(\vartheta) + o_{P_{n\vartheta}}(1) \qquad (7.4)$$

with

$$G_{nt}(\vartheta) = Z_{n\vartheta}(g_{t\vartheta}) = c_n^{-1} \int_0^t U_\vartheta^{-1} Y_n' W_{n\vartheta} \, dM_{n\vartheta}(s) \qquad (7.5)$$

In the following we illustrate three methods to construct regular and efficient estimators for $k_t(\vartheta) = \vartheta(t)'$. In each method, we replace the integrand $U_\vartheta^{-1} Y_n' W_{n\vartheta}$ in (7.5) by an estimator based on an estimator $\tilde{\lambda}_n$ for λ such that

$$\tilde{W}_n = \mathrm{diag}(Y_{ni}\tilde{\lambda}_n)^{-1} \qquad \text{and} \qquad \tilde{U}_n = c_n^{-2}Y_n'\tilde{W}_nY_n$$

fulfill

$$\int (\tilde{U}_n^{-1}Y_n'\tilde{W}_n - U_\vartheta^{-1}Y_n'W_{n\vartheta})\, dM_{n\vartheta}(s) = o_{P_{n\vartheta}}(1) \qquad (7.6)$$

One can use the estimator $\tilde{\lambda}_n$ defined by McKeague (1988) as follows. As a preliminary estimator for ϑ, use the *Aalen estimator* $\tilde{\vartheta}_n$ defined by

$$d\tilde{\vartheta}_n = (Y_n'Y_n)^{-1}Y_n'\, dX_n$$

Set

$$\hat{\lambda}_n(t) = b_n^{-1}\int K(b_n^{-1}(t-s))\, d\tilde{\vartheta}_n(s)$$

with an appropriate kernel K and sequence $b_n \to \infty$. This is a multivariate version of Ramlau-Hansen's (1983) estimator; see McKeague (1988, Section 3.1). Estimate $W_{ni\,\vartheta}$, $W_{n\vartheta}$, U_ϑ by

$$\tilde{W}_{ni} = (Y_{ni}\tilde{\lambda}_n)^{-1} \qquad \tilde{W}_n = \mathrm{diag}\,\tilde{W}_{ni} \qquad \tilde{U}_n = c_n^{-2}Y_n'\tilde{W}_nY_n.$$

Under McKeague's conditions, we can verify that the quadratic variation process of the martingale

$$\int_{t_0}^t (\tilde{U}_n^{-1}Y_n'\tilde{W}_n - U_\vartheta^{-1}Y_n'W_{n\vartheta})\, dM_{n\vartheta}(s)$$

goes to 0 for each $t_0 > 0$ and $t_0 < t \le 1$. The argument is similar to that of McKeague (1988, proof of Theorem 3.2). Under McKeague's conditions, the integrand is bounded in $P_{n\vartheta}$-probability. Hence the quadratic variation process on $[0, t_0]$ can be made small by taking t_0 small. This proves (7.6).

Besides (7.6), we will use relations (7.8) and (7.9) below. Note that for each W,

$$\int_0^t (Y_n'WY_n)^{-1}Y_n'W\, dM_{n\vartheta}(s)$$

$$= \int_0^t (Y_n'WY_n)^{-1}Y_n'W\, dX_n(s) - \vartheta(t) \qquad (7.7)$$

Define

$$\tilde{G}_{nt}(k) = c_n^{-1}\int_0^t \tilde{U}_n^{-1}Y_n'\tilde{W}_n\, dX_n(s) - c_nk \qquad k \in \mathbb{R}^p$$

Relation (7.7), for $W = \tilde{W}_n$, and (7.6) imply

$$\tilde{G}_{nt}(\vartheta(t)) = c_n^{-1} \int_0^t \tilde{U}_n^{-1} Y_n' \tilde{W}_n \, dM_{n\vartheta}(s) + o_{P_{n\vartheta}}(1)$$

$$= G_{nt}(\vartheta) + o_{P_{n\vartheta}}(1) \qquad (7.8)$$

Note that for $k, m \in \mathbb{R}^p$,

$$\tilde{G}_{nt}(m) - \tilde{G}_{nt}(k) + c_n(m - k) = 0 \qquad (7.9)$$

1. The most direct approach to constructing an efficient estimator $\hat{\vartheta}_n(t)$ for $\vartheta(t)$ is the following. By (7.8),

$$G_{nt}(\vartheta) = c_n^{-1} \int_0^t \tilde{U}_n^{-1} Y_n' \tilde{W}_n \, dX_n(s) - c_n \vartheta(t) + o_{P_{n\vartheta}}(1)$$

Hence relation (7.4) is fulfilled by the estimator

$$\hat{\vartheta}_n(t) = c_n^{-2} \int_0^t \tilde{U}_n^{-1} Y_n' \tilde{W}_n \, dX_n(s) \qquad (7.10)$$

This proves that $\hat{\vartheta}_n(t)$ is regular and efficient for $\vartheta(t)$. Introduce

$$\vartheta_n^*(t) = \int_0^t 1\{U_n \text{ invertible}\} \lambda \, ds$$

This is a slight modification of $\vartheta(t)$. The difference between $\hat{\vartheta}_n(t)$ and $\vartheta_n^*(t)$ can be written as a stochastic integral with respect to $M_{n\vartheta}$,

$$\hat{\vartheta}_n(t) - \vartheta_n^*(t) = c_n^{-2} \int_0^t \tilde{U}_n^{-1} Y_n' \tilde{W}_n \, dM_{n\vartheta}(s)$$

In this sense, $\hat{\vartheta}_n(t)$ is a *martingale estimator*.

Let us also illustrate the two other methods for constructing efficient estimators.

2. If we have a preliminary estimator $\tilde{\vartheta}_n(t)$ for $\vartheta(t)$, we can try to use the improvement procedure outlined in Section 5. A candidate for the preliminary estimator is

$$\tilde{\vartheta}_n(t) = \int_0^t (Y_n' Y_n)^{-1} Y_n' \, dX_n(s)$$

This is Aalen's least squares estimator, evaluated at time t. An "estimator" for G_{nt} is $\tilde{G}_{nt}(\tilde{\vartheta}_n(t))$. From (7.9) and (7.8) we obtain relation (5.3):

$$\tilde{G}_{nt}(\tilde{\vartheta}_n(t)) = \tilde{G}_{nt}(\vartheta(t)) - c_n(\tilde{\vartheta}_n(t) - \vartheta(t))$$
$$= G_{nt}(\vartheta) - c_n(\tilde{\vartheta}_n(t) - \vartheta(t)) + o_{P_{n\vartheta}}(1)$$

Hence we can apply (5.4) to get (7.4), and hence regularity and efficiency, for the improved estimator

$$\hat{\vartheta}_n(t) = \tilde{\vartheta}_n(t) + c_n^{-1}\tilde{G}_{nt}(\tilde{\vartheta}_n(t)) \tag{7.11}$$

Note that by definition of $\tilde{G}_{nt}$, improving $\tilde{\vartheta}_n(t)$ reproduces the estimator (7.10):

$$\tilde{\vartheta}_n(t) + c_n^{-1}\tilde{G}_{nt}(\tilde{\vartheta}_n(t)) = c_n^{-2}\int_0^t \tilde{U}_n^{-1}Y_n'\tilde{W}_n \, dX_n(s)$$

3. The estimator $\hat{\vartheta}_n(t)$ defined in (7.10) or (7.11) can also be interpreted as solution of an efficient estimating equation by the reasoning of Section 6. Following Remark 6.7, we extend k_t and G_{nt} to $D = D[0, 1]^p$. The functional $k_t(\vartheta) = \vartheta(t)'$ has a natural extension, evaluation at t. For $t \notin \Theta$ set

$$G_{nt}(\tau) = \tilde{G}_{nt}(\tau(t))$$

Consider the estimator $\hat{\vartheta}_n$ defined by

$$d\hat{\vartheta}_n = c_n^{-2}\tilde{U}_n^{-1}Y_n'\tilde{W}_n \, dX_n \tag{7.12}$$

(Note that $\hat{\vartheta}_n(t)$ equals the estimator of (7.10) or (7.11).) Because $\hat{\vartheta}_n$ is a pure jump process,

$$P_{n\vartheta}\{\hat{\vartheta}_n \in \Theta\} \to 0 \tag{7.13}$$

The estimator solves, simultaneously for each $t \in [0, 1]$, the estimating equation

$$G_{nt}(\hat{\vartheta}_n) = 0$$

Let us check the assumptions of Proposition 6.4. McKeague (1988, p. 144, Theorem 3.2) shows that $c_n(\hat{\vartheta}_n - \vartheta)$ converges in distribution to a p-dimensional continuous Gaussian martingale. This implies tightness (6.2). To apply Proposition 6.4, it remains to check (6.1). By (7.13) and Remark 6.6, it suffices to consider $\tau \notin \Theta$. From (7.8) and (7.9) we obtain uniformly for $\tau \notin \Theta$,

$$G_{nt}(\tau) - G_{nt}(\vartheta) = \tilde{G}_{nt}(\tau(t)) - G_{nt}(\vartheta)$$
$$= \tilde{G}_{nt}(\tau(t)) - \tilde{G}_{nt}(\vartheta(t)) + o_{P_{n\vartheta}}(1)$$
$$= -c_n(\tau(t) - \vartheta(t)) + o_{P_{n\vartheta}}(1)$$

Relation (7.4) now follows from Proposition 6.4, and $\hat{\vartheta}_n$ is regular and efficient.

The estimator $\hat{\vartheta}_n$ defined in (7.12) was introduced by Huffer and McKeague (1990, Section 5.3) as a weighted least squares estimator. McKeague and Utikal (1990, Section 3.2) motivate the choice of weights by interpreting the model as a linear regression model with heteroscedastic errors. The asymptotic distribution of $\hat{\vartheta}_n$ was obtained by McKeague (1988, p. 144, Theorem 3.2). Its asymptotic variance agrees with the variance bound in Greenwood and Wefelmeyer (1990a, Section 3). Hence $\hat{\vartheta}_n$ is efficient. We have seen that efficiency (and regularity) is also implied by each of the three different interpretations of $\hat{\vartheta}_n$ given above.

8 COX'S PROPORTIONAL HAZARDS MODEL

The counting process version of Cox's (1972) *proportional hazards model* was introduced by Andersen and Gill (1982). It is of the form considered in Section 3, with intensity process

$$a_{ni\,\vartheta}(s) = C_{ni}(s)\lambda(s)\exp(\beta'Y_{ni}(s)) \qquad s \in [0, 1] \qquad (8.1)$$

Here C_{ni} is a predictable *censoring process* taking only values 0 and 1, Y_{ni} is a p-dimensional vector of predictable *covariate processes*, β is a p-dimensional vector of *regression coefficients*, and λ is a bounded *baseline hazard function*. Assume for simplicity that the covariate processes Y_{ni} are uniformly bounded.

Parametrize by $\vartheta = (\beta, \Lambda)$, where Λ is the *cumulative baseline hazard rate*, defined by

$$\Lambda(t) = \int_0^t \lambda\,ds$$

Let A denote the set of all such Λ with λ bounded. Then $\Theta = \mathbb{R}^p \times A$. Fix β and Λ. Define

$$\tilde{S}_{0n\gamma} = c_n^{-2}\sum_{i=1}^n C_{ni}\exp(\gamma'Y_{ni}) \qquad (8.2)$$

$$\tilde{S}_{1n\gamma} = c_n^{-2}\sum_{i=1}^n C_{ni}Y_{ni}\exp(\gamma'Y_{ni}) \qquad (8.3)$$

$$\tilde{S}_{2n\gamma} = c_n^{-2} \sum_{i=1}^{n} C_{ni} Y_{ni} Y_{ni}' \exp(\gamma' Y_{ni}) \tag{8.4}$$

Assume that there exist scalar, vector and matrix functions $S_{0\gamma\Lambda}$, $S_{1\gamma\Lambda}$, $S_{2\gamma\Lambda}$, with $S_{0\gamma\Lambda}(s)$ and $S_{1\gamma\Lambda}(s)$ uniformly continuous in $\gamma = \beta$ for $s \in [0, 1]$, and $S_{0\beta\Lambda}$ positive and continuous, such that, for some $\epsilon > 0$,

$$\sup_{|\gamma - \beta| \le \epsilon} \sup_{t \in [0,1]} |\tilde{S}_{in\gamma} - S_{i\gamma\Lambda}| \to 0 \qquad (P_{n\beta\Lambda}) \tag{8.5}$$

Assume that

$$\Sigma_{\beta\Lambda} = \int (S_{2\beta\Lambda} - S_{0\beta\Lambda}^{-1} S_{1\beta\Lambda} S_{1\beta\Lambda}') \lambda \, ds \quad \text{is positive definite} \tag{8.6}$$

is positive definite.

We will apply the arguments of Sections 5 and 6 to construct efficient estimators for β. The treatment parallels the one for Aalen's additive risk model in Section 7.

We introduce a local model at $\vartheta = (\beta, \Lambda)$ as follows. Let B be the set of all bounded functions on $[0, 1]$. Set $V = \mathbb{R}^p \times B$. For $(b, v) \in V$ define $\vartheta_{nbv} = (\beta_{nb}, \Lambda_{nv})$ by

$$\beta_{nb} = \beta + c_n^{-1} b \tag{8.7}$$

$$d\Lambda_{nv} = (1 + c_n^{-1} v) \, d\Lambda \tag{8.8}$$

According to Greenwood and Wefelmeyer (1990a, Section 4), the (partially specified) log-likelihood (3.2) is *locally asymptotically normal* in the following sense:

$$L_{n\vartheta\vartheta_{nbv}} = Z_{n\vartheta}(b, v) - \tfrac{1}{2}\|(b, v)\|_\vartheta^2 + o_{P_{n\vartheta}}(1)$$

Here the linear process $Z_{n\vartheta}$ is

$$Z_{n\vartheta}(b, v) = c_n^{-1} \sum_{i=1}^{n} \int (b' Y_{ni} + v) \, dM_{ni\,\vartheta}(s)$$

with the martingale $M_{ni\,\vartheta}$ defined by

$$dM_{ni\,\vartheta}(s) = dX_{ni}(s) - C_{ni} \lambda \exp(\beta' Y_{ni}) \, ds$$

The norm $\| \; \|_\vartheta$ is defined by

$$\|(b, v)\|_\vartheta^2 = \int (b' S_{2\vartheta} b + 2b' S_{1\vartheta} v + S_{0\vartheta} v^2) \lambda \, ds$$

The *acuity* is defined as the inner product corresponding to this norm,

$$((b, v), (c, w))_{\vartheta} = \int (b'S_{2\vartheta}c + b'S_{1\vartheta}w + vS'_{1\vartheta}c + vS_{0\vartheta}w)\lambda\,ds \quad (8.9)$$

Let us now consider the problem of finding an efficient estimator for the vector β of regression coefficients. Writing the regression coefficients as a *row* vector for convenience, we introduce the p-dimensional functional

$$k(\beta, \Lambda) = \beta'$$

Note that the functional is just a projection and hence linear. We have

$$c_n(k(\vartheta_{nbv}) - k(\vartheta)) = c_n(\beta_{nb} - \beta)' = b'$$

Following Greenwood and Wefelmeyer (1990a, Section 4), we express the linear functional

$$(b, v) \to b'$$

on the local parameter space $V = \mathbb{R}^p \times B$ in terms of the acuity (8.9):

$$b' = ((b, v), (b_{\vartheta}, v_{\vartheta}))_{\vartheta} \qquad \text{for } (b, v) \in V$$

with *gradient* $(b_{\vartheta}, v_{\vartheta})$ given by

$$b_{\vartheta} = \Sigma_{\vartheta}^{-1} \qquad v_{\vartheta} = -S_{0\vartheta}^{-1}S'_{1\vartheta}\Sigma_{\vartheta}^{-1}$$

As noted in Remark 5.5, a regular and efficient estimator $\hat{\beta}_n$ for β is characterized by

$$c_n(\hat{\beta}_n - \beta)' = G_n(\vartheta) + o_{P_{n\vartheta}}(1) \quad (8.10)$$

with

$$G_n(\vartheta) = Z_{n\vartheta}(b_{\vartheta}, v_{\vartheta})$$

$$= \Sigma_{\vartheta}^{-1}c_n^{-1}\sum \int (Y_{ni} - S_{0\vartheta}^{-1}S_{1\vartheta})\,dM_{ni\,\vartheta}(s) \quad (8.11)$$

In the following we illustrate three methods to construct regular and efficient estimators for β. In each method, we replace $S_{0\vartheta}^{-1}S_{1\vartheta} = S_{0\beta\Lambda}^{-1}S_{1\beta\Lambda}$ by $\tilde{S}_{0n\beta}^{-1}\tilde{S}_{1n\beta}$, with $\tilde{S}_{0n\beta}$ and $\tilde{S}_{1n\beta}$ defined in (8.2) and (8.3), respectively. Let us first derive some relations ((8.13), (8.15), (8.18)) which will be used below. By (8.5),

$$c_n^{-1}\sum \int (\tilde{S}_{0n\beta}^{-1}\tilde{S}_{1n\beta} - S_{0\beta\Lambda}^{-1}S_{1\beta\Lambda})\,dM_{ni\beta\Lambda}(s) = o_{P_{n\beta\Lambda}}(1) \quad (8.12)$$

From (8.1) and the definitions of $\tilde{S}_{0n\beta}$ and $\tilde{S}_{1n\beta}$ we obtain

$$c_n^{-1} \sum \int (Y_{ni} - \tilde{S}_{0n\beta}^{-1}\tilde{S}_{1n\beta}) \, dM_{ni\beta\Lambda}(s)$$

$$= c_n^{-1} \sum \int (Y_{ni} - \tilde{S}_{0n\beta}^{-1}\tilde{S}_{1n\beta}) \, dX_{ni}(s) + o_{P_{n\beta\Lambda}}(1) \qquad (8.13)$$

Let $\tilde{\Sigma}_n$ be a consistent estimator for $\Sigma_{\beta\Lambda}$. Define

$$\tilde{G}_n(\beta) = \tilde{\Sigma}_n^{-1} c_n^{-1} \sum \int (Y_{ni} - \tilde{S}_{0n\beta}^{-1}\tilde{S}_{1n\beta}) \, dX_{ni}(s) \qquad (8.14)$$

Relations (8.12) and (8.13) imply

$$\tilde{G}_n(\beta) = G_n(\beta, \Lambda) + o_{P_{n\beta\Lambda}}(1) \qquad (8.15)$$

It is easily seen from the continuity assumptions on $S_{i\gamma\Lambda}$ and uniform boundedness of the covariates that uniformly for $c_n(\gamma - \beta)$ bounded,

$$c_n^{-1} \sum \int (\tilde{S}_{0n\gamma}^{-1}\tilde{S}_{1n\gamma} - \tilde{S}_{0n\beta}^{-1}\tilde{S}_{1n\beta}) \, dM_{ni\gamma\Lambda} = o_{P_{n\beta\Lambda}}(1) \qquad (8.16)$$

and, by a Taylor expansion of the exponential function,

$$c_n^{-1} \sum \int (Y_{ni} - \tilde{S}_{0n\beta}^{-1}\tilde{S}_{1n\beta})C_{ni}(\exp(\gamma'Y_{ni}) - \exp(\beta'Y_{ni}))\lambda \, ds$$

$$= c_n^{-1} \sum \int (Y_{ni} - \tilde{S}_{0n\beta}^{-1}\tilde{S}_{1n\beta})C_{ni}Y_{ni}'\exp(\beta'Y_{ni})\lambda \, ds(\gamma - \beta) + o_{P_{n\beta\Lambda}}(1)$$

$$= \int (\tilde{S}_{2n\beta} - \tilde{S}_{0n\beta}^{-1}\tilde{S}_{1n\beta}\tilde{S}_{1n\beta}')\lambda \, ds c_n(\gamma - \beta) + o_{P_{n\beta\Lambda}}(1)$$

$$= \Sigma_{\beta\Lambda} c_n(\gamma - \beta) + o_{P_{n\beta\Lambda}}(1) \qquad (8.17)$$

Relations (8.13), (8.16) and (8.17), together with consistency of $\tilde{\Sigma}_n$, imply uniformly for $c_n(\gamma - \beta)$ bounded,

$$\tilde{G}_n(\gamma) - \tilde{G}_n(\beta)$$

$$= \tilde{\Sigma}_n^{-1} c_n^{-1} \sum \int (Y_{ni} - \tilde{S}_{0n\gamma}^{-1}\tilde{S}_{1n\gamma}) \, dM_{ni\gamma\Lambda}(s)$$

$$- \tilde{\Sigma}_n^{-1} c_n^{-1} \sum \int (Y_{ni} - \tilde{S}_{0n\beta}^{-1}\tilde{S}_{1n\beta}) \, dM_{ni\beta\Lambda}(s) + o_{P_{n\beta\Lambda}}(1)$$

$$= -\tilde{\Sigma}_n^{-1} \int (\tilde{S}_{2n\beta} - \tilde{S}_{0n\beta}^{-1}\tilde{S}_{1n\beta}\tilde{S}_{1n\beta}')\lambda \, dsc_n(\gamma - \beta) + o_{P_{n\beta\Lambda}}(1)$$

$$= -c_n(\gamma - \beta) + o_{P_{n\beta\Lambda}}(1) \tag{8.18}$$

1. As shown by Andersen and Gill (1982), the Cox maximum partial likelihood estimator $\hat{\beta}_n$ for β is c_n-consistent and fulfills

$$\tilde{G}_n(\hat{\beta}_n) = 0 \tag{8.19}$$

From (8.18), (8.19) and (8.15) we obtain

$$c_n(\hat{\beta}_n - \beta) = \tilde{G}_n(\beta) + o_{P_{n\beta\Lambda}}(1)$$
$$= G_n(\beta, \Lambda) + o_{P_{n\beta\Lambda}}(1)$$

This is relation (8.10). Hence $\hat{\beta}_n$ is regular and efficient for β.

Let us also illustrate the other two methods for constructing efficient estimators.

2. If we have a preliminary c_n-consistent estimator $\tilde{\beta}_n$ for β, we can try to use the improvement procedure outlined in Section 5. An estimator for G_n is $\tilde{G}_n(\tilde{\beta}_n)$. From (8.18) and (8.15) we obtain relation (5.3):

$$\tilde{G}_n(\tilde{\beta}_n) = \tilde{G}_n(\beta) - c_n(\tilde{\beta}_n - \beta) + o_{P_{n\beta\Lambda}}(1)$$
$$= G_n(\beta, \Lambda) - c_n(\tilde{\beta}_n - \beta) + o_{P_{n\beta\Lambda}}(1)$$

Hence we can apply (5.4) to get (8.10), and therefore regularity and efficiency, for the improved estimator

$$\hat{\beta}_n = \tilde{\beta}_n + c_n^{-1}\tilde{G}_n(\tilde{\beta}_n)$$

3. The Cox estimator $\hat{\beta}_n$ can also be interpreted as solution of an efficient estimating equation by the reasoning of Section 6. Following Remark 6.7, we extend $k(\beta, \Lambda) = \beta'$ and G_n from $\Theta = \mathbb{R}^p \times A$ to $D = \mathbb{R}^p \times D[0, 1]$. The functional k has a natural extension. For $\gamma \in \mathbb{R}^p$ and $M \notin A$ set

$$G_n(\gamma, M) = \tilde{G}_n(\gamma)$$

Introduce a dummy estimator $\tilde{\Lambda}_n$ such that

$$P_{n\beta\Lambda}\{\tilde{\Lambda}_n \in A\} = 0$$

Then

$$P_{n\beta\Lambda}\{(\hat{\beta}_n, \tilde{\Lambda}_n) \in \Theta\} = 0 \tag{8.20}$$

It follows from (8.19) that $(\hat{\beta}_n, \tilde{\Lambda}_n)$ solves the estimating equation

$$G_n(\hat{\beta}_n, \tilde{\Lambda}_n) = 0 \qquad (8.21)$$

Let us check the assumptions of Proposition 6.4. Since $G_n(\gamma, M)$ is constant for $M \notin A$, it suffices to have c_n-consistency (6.2) for $\hat{\beta}_n$ in place of $(\hat{\beta}_n, \tilde{\Lambda}_n)$. By (8.20) and Remark 6.6, it suffices to check (6.1) for $(\gamma, M) \notin \Theta$. From (8.15) and (8.18) we obtain uniformly for $(\gamma, M) \notin \Theta$,

$$\begin{aligned}
G_n(\gamma, M) - G_n(\beta, \Lambda) &= \tilde{G}_n(\gamma) - G_n(\beta, \Lambda) \\
&= \tilde{G}_n(\gamma) - \tilde{G}_n(\beta) + o_{P_{n\beta\Lambda}}(1) \\
&= -c_n(\gamma - \beta) + o_{P_{n\beta\Lambda}}(1)
\end{aligned}$$

Relation (8.10) now follows from Proposition 6.4, and $\hat{\beta}_n$ is regular and efficient for β.

The asymptotic distribution of the Cox estimator $\hat{\beta}_n$ was obtained by Andersen and Gill (1982, p. 1106, Theorem 3.2). Its asymptotic covariance agrees with the covariance bound in Greenwood and Wefelmeyer (1990a, Section 4). Hence $\hat{\beta}_n$ is efficient. We have seen that efficiency (and regularity) is also implied by each of the two different interpretations of $\hat{\beta}_n$ given above. Dzhaparidze (1985) proves efficiency of $\hat{\beta}_n$ for the case that the filtration is generated by the counting process. For the Tsiatis (1981) version of the Cox model, with time-independent covariates, efficiency of $\hat{\beta}_n$ was already proved by Begun et al. (1983, p. 448, Example 4).

From Andersen and Gill (1982, p. 1103) we see that $\tilde{G}_n(\beta)$ equals the derivative of the Cox partial likelihood. By (8.15) this derivative is close to the derivative of the (partially specified) likelihood in the least favorable direction for β. This explains why the Cox estimator is efficient.

Here we have interpreted nonparametric maximum likelihood estimators as solutions of estimating equations based on an extension of the *derivatives* of the likelihood. This is the approach suggested by Gill (1989). The Cox estimator can also be seen to maximize an appropriate extension of the likelihood *itself*. Johansen (1983, p. 166) indicates that the Cox estimator maximizes the likelihood profile for a certain extension. Jacobsen (1984, p. 204) indicates that the likelihood is maximized approximately for a different extension. For the case of time-independent covariates, Bailey (1984, p. 732, Theorem 1) shows that the Cox estimator is asymptotically equivalent to the first component of the estimator maximizing an extended likelihood.

ACKNOWLEDGMENT

This work was begun while the authors were visiting the Department of Mathematical Sciences at The Johns Hopkins University. We are most grateful to Robert J. Serfling for his kind hospitality. The results were presented at an Oberwolfach meeting on martingale methods in statistics in December 1988. The paper was finished while the second author was visiting the Department of Mathematics at the University of British Columbia, Vancouver. We thank Richard Gill, Jean Jacod, and Ian McKeague for access to unpublished manuscripts, and the referee for a number of comments.

REFERENCES

Aalen, O. O. (1975). *Statistical Inference for a Family of Counting Processes.* Ph.D. dissertation, University of California, Berkeley.

Aalen, O. O. (1978) Nonparametric inference for a family of counting processes. *Ann. Statist.*, 6, 701–725.

Aalen, O. O. (1980). A model for nonparametric regression analysis of counting processes. In *Mathematical Statistics and Probability Theory* (W. Klonecki, A. Kozek and J. Rosiński, eds.), 1–25, Springer-Verlag, New York.

Andersen, P. K. and Gill, R. D. (1982). Cox's regression model for counting processes: a large sample study. *Ann. Statist.*, 10, 1100–1120.

Arjas, E. and Haara, P. (1984). A marked point process approach to censored failure data with complicated covariates. *Scand. J. Statist.*, 11, 193–209.

Bailey, K. R. (1984). Asymptotic equivalence between the Cox estimator and the general ML estimators of regression and survival parameters in the Cox model. *Ann. Statist.*, 12, 730–736.

Basawa, I. V., and Koul, H. L. (1988). Large sample statistics based on quadratic dispersion. *Int. Statist. Rev.*, 56, 199–219.

Begun, J. M., Hall, W. J., Huang, W.-M., and Wellner, J. A. (1983). Information and asymptotic efficiency in parametric-nonparametric models. *Ann. Statist.*, 11, 432–452.

Bickel, P. J. (1982). On adaptive estimation. *Ann. Statist.*, 10, 647–671.

Cox, D. R. (1972). Regression models and life tables (with discussion). *J. Roy. Statist. Soc. Ser. B*, 34, 187–220.

Dzhaparidze, K. (1985). On asymptotic inference about intensity parameters of a counting process. *Bull. Int. Statist. Inst.*, *51*, Vol. 4, 23.2–1-23.3–15.

Gill, R. D. (1985). Notes on product integration, likelihood and partial likelihood for counting processes, noninformative and independent censoring. Unpublished manuscript.

Gill, R. D. (1987). Non- and semiparametric maximum likelihood estimators and the von Mises method (Part 2). Unpublished manuscript.

Gill, R. D. (1989). Non- and semiparametric maximum likelihood estimators and the von Mises method (Part 1). *Scand. J. Statist.*, *16*, 97–128.

Greenwood, P. E. (1988). Partially specified semimartingale experiments. In *Statistical Inference from Stochastic Processes* (N. U. Prabhu, ed.), 1–17, *Contemporary Math.*, *80*, Amer. Math. Soc.

Greenwood, P. E. and Shiryaev, A. N. (1990). Sequential estimation for first order autoregressive models. To appear in *Stochastics*.

Greenwood, P. E. and Wefelmeyer, W. (1990a). Efficiency of estimators for partially specified filtered models. To appear in *Stoch. Proc. Appl.*

Greenwood, P. E. and Wefelmeyer, W. (1990b). Efficient estimation in a nonlinear counting process regression model. To appear in *Canad. J. Statist.*

Huffer, F. W. and McKeague, I. W. (1990). Survival analysis using additive risk models. To appear in *J. Amer. Statist. Assoc.*

Jacobsen, M. (1984). Maximum likelihood estimation in the multiplicative intensity model: a survey. *Int. Statist. Rev.*, *52*, 193–207.

Jacod, J. (1975). Multivariate point processes: predictable projection, Radon-Nikodym derivatives, representation of martingales. *Z. Wahrscheinlichkeitstheorie verw. Gebiete*, *31*, 235–253.

Jacod, J. (1987). Partial likelihood process and asymptotic normality. *Stoch. Proc. Appl.*, *26*, 47–71.

Jacod, J. (1989). Une application de la topologie d'Emery: le processus information d'une modèle statistique filtré. In *Séminaire de Probabilités* XXIII (J. Azéma, P. A. Meyer, M. Yor, eds.), 448–474, Lecture Notes in Mathematics 1372, Springer-Verlag, Berlin.

Jacod, J. (1990). Sur le processus de vraisemblance partielle. *Ann. Inst. Henri Poincaré, Probab. Statist.*, *26*, 299–329.

Jacod, J. and Shiryaev, A. N. (1987). *Limit Theorems for Stochastic Processes*. Grundlehren der mathematischen Wissenschaften 288. Springer-Verlag, Berlin.

Johansen, S. (1983). An extension of Cox's regression model. *Int. Statist. Rev.*, *51*, 165–174.

Klaassen, C. A. J. (1987). Consistent estimation of the influence function of locally asymptotically linear estimators. *Ann. Statist.*, *15*, 1548–1562.

Le Cam, L. (1986). *Asymptotic Methods in Statistical Decision Theory*. Springer Series in Statistics. Springer-Verlag, New York.

McKeague, I. W. (1988). Asymptotic theory for weighted least squares estimators in Aalen's additive risk model. In *Statistical Inference from Stochastic Processes* (N. U. Prabhu, ed.), 139–152, *Contemporary Math.*, *80*, Amer. Math. Soc.

McKeague, I. W. and Utikal, K. J. (1990). Stochastic calculus as a tool in survival analysis. To appear in *Appl. Math. Comput.*

Ramlau-Hansen, H. (1983). Smoothing counting process intensities by means of kernel functions. *Ann. Statist.*, *11*, 453–466.

Schick, A. (1987). A note on the construction of asymptotically linear estimators. *J. Statist. Plann. Inference*, *16*, 89–105. Correction 22 (1989) 269–270.

Slud, E. (1986). Martingale Methods in Statistics. Book manuscript.

Tsiatis, A. A. (1981). A large sample study of Cox's regression model. *Ann. Statist.*, *9*, 93–108.

5

Nonparametric Estimation of Trends in Linear Stochastic Systems

Ian W. McKeague*
Florida State University, Tallahassee, Florida

Tiziano Tofoni
Postgraduate School of Telecommunications, L'Aquila, Italy

Techniques for the estimation of unknown additive trends present in the state and measurement processes of a Kalman–Bucy linear system are introduced. We obtain asymptotic results describing the performance of the estimators under i.i.d. and periodic observation schemes. The observed process is given by $dY(t) = g(t)\,dt + dZ(t)$, where Z is the measurement process and g is an unknown trend function, and there is an additive trend f present in the state process X. The problem is to estimate f and g, and remove them from the measurement process. Trend removal involves replacing f and g in the Kalman filter $\tilde{X}(t) = E(X(t) \mid \mathcal{F}_t^Y)$ — based on observation of Y — by appropriate estimates. We show that this can be done under the following observation schemes: (I) n i.i.d. replicate of Y over a fixed interval $[0, T]$, (II) observation of a single trajectory of Y over a long interval $[0, nT]$, where f, g and the functions defining the linear system are periodic with period T.

---

*Research supported by Army Research Office Grant DAAL03-86-K-0094.

143

1 INTRODUCTION

Consider a linear stochastic system of the type introduced by Kalman and Bucy (1961): A p-dimensional "state" process X and a q-dimensional "measurement" process Z are given by the stochastic differential equations

$$dX(t) = A(t)X(t) \, dt + B(t)u(t) \, dt + dW(t)$$
$$dZ(t) = C(t)X(t) \, dt + dV(t)$$

$0 \leq t \leq T$, where W and V are independent p and q-dimensional Wiener processes, $u(\cdot)$ is a known deterministic input, A, B, C are known nonrandom time-varying matrices of suitable dimensions, $X(0)$ is independent of W and V, the mean $E(X(0)) = m$ and covariance matrix of $X(0)$ are known, and $Z(0) = 0$. The Kalman filtering theory provides recursive formulae for the conditional expectation $\hat{X}(t) = E(X(t) \mid \mathcal{F}_t^Z)$ which is the optimal mean square estimate of the state $X(t)$ given the past $\mathcal{F}_t^Z = \sigma(Z_s, 0 \leq s \leq t)$ of the measurement process, see Liptser and Shiryayev (1978) and Kallianpur (1980).

In real applications of the Kalman filter to signal processing it is often found that unknown additive trends are present in the state and measurement processes; that is, the state process X is given by

$$dX(t) = f(t) \, dt + A(t)X(t) \, dt + B(t)u(t) \, dt + dW(t) \qquad (1)$$

and instead of observing Z, we observe the process Y given by

$$dY(t) = g(t) \, dt + dZ(t) \qquad Y(0) = 0 \qquad (2)$$

where f and g are unknown "trend" functions.

In the present paper we shall consider the problem of estimating the trends f and g and removing them from the measurement process. Trend removal amounts to replacing the functions f and g used in the Kalman filter $\hat{X}(t) = E(X(t) \mid \mathcal{F}_t^Y)$—based on observation of Y—by appropriate estimates $\hat{f}$ and $\hat{g}$.

Two types of observation scheme are considered:

(I) n realizations $\{Y_i(t), t \in [0, T], i = 1, \ldots, n\}$ of the process Y satisfying (1) and (2) with the corresponding system realizations having independent noise processes W_i and V_i, $i = 1, \ldots, n$.

(II) observation of a single trajectory of Y over the interval $[0, nT]$, where the functions f, g, A, B and C are periodic with period T.

Observation scheme (II) is relevant to situations where there is a "time-of-day" or "seasonal" effect present in the model; for example, in the analysis of circadian rhythm data in biology, or in the study of cyclic systems in control engineering — see the review article of Bittanti and Guardabassi (1986). We are interested in the asymptotic properties of estimators of f and g as $n \to \infty$ with T remaining fixed. We shall see that f and g are not identifiable unless one of them is absent from the model (that is, $f = 0$, $g \neq 0$ or $f \neq 0$, $g = 0$).

There is a vast literature on the estimation of finite dimensional parameters in discrete time linear stochastic systems; refer to the books of Davis and Vinter (1984) and Kumar and Varaiya (1986). In continuous time such problems were first studied by Balakrishnan (1973). Further contributions have been made by Bagchi (1980), Tugnait (1980) and Bagchi and Borkar (1984). Nonparametric estimation for linear stochastic systems is considered to be a difficult problem; see, for instance, the closing comment of a recent paper of Aihara and Bagchi (1989). In general the functions A, C, f and g are not even identifiable. In the present paper we are studying the very special case in which A, B and C are known, and at least one of the trend functions is known to be absent.

There is an extensive literature on nonparametric estimation for the drift (or trend) function, g, in a diffusion process satisfying (2) with Z as a Wiener process; see Ibragimov and Khasminski (1980, 1981), Geman and Hwang (1983), Nguyen and Pham (1982), Beder (1987), McKeague (1986) — who allowed Z to be a general square integrable martingale, and Leskow (1989) — who considered the case of a periodic model. These authors use either Parzen–Rosenblatt type kernel estimators or Grenander (1980) sieve estimators for g, but those estimators are not directly applicable to the present setting, unless C is identically zero (in which case only g is identifiable). We shall find that there is a function h, related to g and f through two Volterra integral equations, and h can be estimated by kernel or sieve type estimators. Estimates of g and f can then be obtained by inserting estimates of h or its first derivative h' in the solutions of the Volterra integral equations.

The paper is organized as follows. Section 2 contains introductory discussion concerning the basic innovations representation of the observation process, identifiability, bias under misspecified trends, and schemes (I) and (II). Estimation of the trend in the measurement process under schemes (I) and (II) is treated in Sections 3 and 4 respectively. In Section 5 we consider estimation of the trend in the state process. In these sections, to simplify the presentation, we assume that the state and measurement processes are one-dimensional ($p = q = 1$). Section 6 contains remarks on the multidimensional case. In Section 7 we indicate some directions for further work.

To conclude this section we shall briefly put our problem in perspective with other inference problems for stochastic processes. Statistical models for stochastic processes are of two broad types. If we observe a process $Y = (Y_t, t \geq 0)$ and we have a covariate process $X = (X_t, t \geq 0)$ to incorporate into the analysis, then we may consider a *partially specified* model in which, loosely speaking (see Greenwood (1988) for a more precise definition), only the conditional distribution of Y given X is specified in terms of an unknown parameter θ. Alternatively, we may know the full joint distribution of (Y, X) for each θ, in which case we have a *fully specified* model. Partially specified models are especially useful and widely applied in the analysis of life history data by taking Y as a counting process describing the times of events in the life of an individual, and X representing a covariate process specific to the individual — the structure of the marginal distribution of X being unspecified; see Arjas and Haara (1984) and Andersen et al. (1988). Fully specified models on the other hand are widely used in the engineering sciences where precise models for the covariate process X can often be developed from well-understood system dynamics. Our model is of this latter type.

Observation schemes may similarly be classified into two broad types: partial and full. In the survival analysis setting partial observation may arise from censoring, truncation, or grouping of the data, see Andersen et al. (1988) and McKeague (1988). In arises in the stochastic systems setting when the state of the system is observed in the presence of noise, as in (1). Despite the diverse applications of such schemes and models, there is a surprising unity to the techniques used. For example, our kernel function techniques are similar to the methods used by Ramlau-Hansen (1983) for the estimation of counting process intensities, and our approach to the periodic case in some ways resembles that of Pons and de Turckheim (1988) to Cox's periodic regression model.

2 THE INNOVATIONS REPRESENTATION

We shall assume throughout that the functions f, g, A, B, C and u are smooth, and $C(t)$ does not vanish anywhere on $[0, T]$. The equations for the Kalman–Bucy filter (see Kallianpur 1980, Section 10.3) are

$$d\hat{X}(t) = [f(t) + A(t)\hat{X}(t) + B(t)u(t)] \, dt + D(t) \, d\nu(t)$$
$$d\nu(t) = dY(t) - [g(t) + C(t)\hat{X}(t)] \, dt \tag{3}$$

where $\hat{X}(0) = m$. The process ν is the so-called *innovations process* which is known to be a standard Wiener process. The function D is the *Kalman gain* which in the present set-up does not depend on f or g. In fact $D(t) = C(t)P(t)$, where P is the unique positive solution to the *Riccati differential equation*

$$P'(t) = 2A(t)P(t) - C^2(t)P^2(t) + 1 \tag{4}$$

with initial condition $P(0) = \mathrm{Var}(X(0))$. From (3) we have

$$d\hat{X}(t) = [A(t) - D(t)C(t)]\hat{X}(t) \, dt + [f(t) + B(t)u(t)$$
$$- D(t)g(t)] \, dt + D(t) \, dY(t)$$

Using Theorem 4.2.4 of Davis (1977) we can solve this equation for $\hat{X}$. Substituting the solution into the second equation in (3) we obtain the following *innovations representation* for Y:

$$Y(t) = \int_0^t [h(s) + U(s)] \, ds + \nu(t) \tag{5}$$

where

$$h(t) = g(t) + C(t) \int_0^t \Psi(t, s)[f(s) - D(s)g(s)] \, ds$$

Here $\Psi(t, s)$ is the solution to the linear time-varying system

$$\frac{\partial \Phi(t, s)}{\partial t} = [A(t) - D(t)C(t)]\Psi(t, s) \qquad \Psi(s, s) = 1$$

and U is given by

$$U(t) = C(t) \left\{ \Psi(t, 0)m + \int_0^t \Psi(t, s)[B(s)u(s) \, ds + D(s) \, dY(s)] \right\}$$

The representation (5) will be of prime importance in the next section.

Identifiability of *f* and *g*

We see from (5) that the function h is identifiable given observation of Y and U; however, f and g are identifiable only in so far as they are uniquely determined in terms of h through (6). Thus, the functions f and g are not in general simultaneously identifiable from observation of Y. However, if the trend is absent from the measurement process ($g = 0$) then (6) reduces to

$$h(t) = \int_0^t \Phi(t, s)f(s)\, ds \qquad (7)$$

where $\Phi(t, s) = C(t)\Psi(t, s)$. If the trend is absent from the state process ($f = 0$) then (6) reduces to

$$h(t) = g(t) + \int_0^t \Gamma(t, s)g(s)\, ds \qquad (8)$$

where $\Gamma(t, s) = -C(t)\Psi(t, s)D(s)$.

As equations involving the unknown f and g, (7) and (8) are *linear Volterra integral equations* of the *first* and *second* kind respectively. It follows from standard results on Volterra equations (see Linz, 1985) that (8) has a unique solution for g, and (7) has a unique solution for f provided $C(t)$ does not vanish on $[0, T]$. Since h is identifiable, the trend f is identifiable when $g = 0$, and g is identifiable when $f = 0$.

2.2 The Log-Likelihood Function

The innovations representation (5) allows us to write down an explicit expression for the log-likelihood function $L(h) = \log[d\mu_h/d\mu_W](Y)$, where μ_h is the measure induced on $C[0, T]$ by Y, and μ_W is Wiener measure. By Lipster and Shiryayev (1977, Theorem 7.7) we have that $\mu_h \ll \mu_W$ and

$$L(h) = \int_0^T \pi(s)\, dY(s) - \frac{1}{2}\int_0^T \pi^2(s)\, ds \qquad (9)$$

where $\pi(s)$ is the term inside the square brackets in (5).

2.3 The Bias Caused by Misspecified Trends

What is the effect on the mean square error of the Kalman filter (3) of using incorrect trend functions $f^* \neq f$, $g^* \neq g$? The answer to this question should provide us with a *modus operandi* for choosing estimators $\hat{f}$, $\hat{g}$ to be used in place of the unknown f, g. Let $\hat{X}^{f,g}(t)$ denote the Kalman filter estimate of $X(t)$ based on (3). The *bias* caused by using f^*, g^* instead of f, g at time t,

$$BIAS(f^*, g^*, t) \equiv \hat{X}^{f^*, g^*}(t) - \hat{X}^{f, g}(t)$$

can be found from (3), cf. Jazwinski (1970, p. 252).

$$BIAS(f^*, g^*, t) = \int_0^t \Psi(t, s)[f^*(s) - f(s) + D(s)(g(s) - g^*(s))]\, ds$$

The increase in the mean square error caused by using estimators f^*, g^* instead of f, g is solely due to this nonrandom bias and is given by $[BIAS(f^*, g^*, t)]^2$.

2.4 Observation Scheme (I)

The processes associated with the ith realization are given the subscript i, as in v_i, u_i, $L_i(h)$ etc. Note that although the observed processes $\{Y_i, i = 1, \ldots, n\}$ are independent, they are not necessarily identically distributed since the inputs u_i are not assumed to be identical for each i. However, the innovations processes v_i are i.i.d. Wiener processes. From (5) we have

$$Y_i(t) = \int_0^t [h(s) + U_i(s)]\, ds + v_i(t) \tag{10}$$

where

$$U_i(t) = C(t)\left\{\Phi(t, 0)m + \int_0^t \Phi(t, s)[B(s)u_i(s)\, ds + D(s)\, dY_i(s)]\right\}$$

The log-likelihood function $L^{(n)}(h)$ is given by

$$L^{(n)}(h) = \sum_{i=1}^n L_i(h)$$

2.5 Observation Scheme (II)

Scheme (II) can be treated using a similar framework to scheme (I). Let h_i, U_i, Y_i and v_i be the following restrictions of h, U, Y and v to the i-th period:

$$h_i(t) = h(iT + t)$$
$$U_i(t) = U(iT + t)$$
$$Y_i(t) = Y(iT + t) - Y(iT)$$
$$v_i(t) = v(iT + t) - v(iT)$$

$0 \le t \le T$. These processes satisfy

$$Y_i(t) = \int_0^t [h_i(s) + U_i(s)] \, ds + v_i(t) \tag{10'}$$

Since v is a Wiener process (which has stationary independent increments), the processes v_i, $i = 1, \ldots, n$ are also i.i.d. Wiener processes on $[0, T]$. The log-likelihood function $L^{(n)}(h)$, given by (9) with T replaced by nT, can be written as

$$L^{(n)}(h) = \sum_{i=1}^{n} L_i(h_i)$$

Note that the function h is not periodic so that $h_i \ne h$. This is the basic difference between schemes (I) and (II), making (II) much harder to analyze.

3 TREND IN THE MEASUREMENT PROCESS – i.i.d. CASE

In this section we consider estimation of g under scheme (I) with $f \equiv 0$. First we introduce estimators $\hat{g}$ of g such that $BIAS(0, \hat{g}, t) \to 0$ uniformly in t as $n \to \infty$ a.s.. In fact, *a fortiori*, $\hat{g}$ will be shown to be strongly L^2-consistent in the sense that $\|g - \hat{g}\| \xrightarrow{a.s.} 0$ as $n \to \infty$, where $\|\cdot\|$ denotes the norm in $L^2[0, T]$.

The basic idea is to take as an estimator of g the solution $\hat{g}$ of the Volterra integral equation

$$\hat{h}(t) = \hat{g}(t) + \int_0^t \Gamma(t, s)\hat{g}(s) \, ds \tag{11}$$

where $\hat{h}$ is an estimator of h. Note that the estimator so obtained is well defined since (11) admits a unique solution whenever $\hat{h} \in L^2[0, T]$ (see Davis, 1977, p. 125). Moreover, should $\hat{h}$ be a strongly L^2-consistent estimator of h, the following theorem shows that $\hat{g}$ is also strongly L^2-consistent.

Theorem 1.   Let $\hat{h}$ be a strongly L^2-consistent estimator of h. Then the solution of the Volterra integral equation (11) is a strongly L^2-consistent estimator of g.

Proof: Let $\bar{h} = \hat{h} - h$ and $\bar{g} = \hat{g} - g$ be the estimation errors of $\hat{h}$ and $\hat{g}$ respectively and denote $M = \sup_{t,s\in[0,T]}|\Gamma(t,s)| < \infty$. Now from (8) and (11)

$$\bar{g}(t) = \bar{h}(t) - \int_0^t \Gamma(t,s)\bar{g}(s)\, ds$$

so that

$$|\bar{g}(t)|^2 \leq 2TM^2 \int_0^t |\bar{g}(s)|^2\, ds + 2|\bar{h}(t)|^2$$

Using Gronwall's inequality (see Kallianpur, 1980, p. 94) we then have

$$|\bar{g}(t)|^2 \leq 4TM^2 \int_0^t |\bar{h}(s)|^2 \exp[2TM^2(t-s)]\, ds + 2|\bar{h}(t)|^2$$

$$\leq 4TM^2 \exp[2T^2M^2] \int_0^t |\bar{h}(s)|^2\, ds + 2|\bar{h}(t)|^2$$

Integrating this last inequality over the interval $[0, T]$ we easily get

$$\|\bar{g}\|^2 \leq (2 + 4T^2M^2 \exp[2T^2M^2])\|\bar{h}\|^2$$

This completes the proof.

3.1 Orthogonal Series Sieve Estimators for *h*

The maximum of the log-likelihood function $L^{(n)}(h)$ is not attained when we maximize over the whole parameter space $L^2[0, T]$. The problem is that the parameter space is too large for the existence of the unconstrained maximum likelihood estimator. One remedy is to apply the *method of sieves* which consists in maximizing the log-likelihood function over an increasing sequence of subsets S_n, $n = 1, 2, \ldots$ of the parameter space. We shall use an orthogonal series sieve $S_n = \text{span}\{\psi_r, r = 1, \ldots, d_n\}$, where $\{\psi_r, r \geq 1\}$ is a complete orthonormal sequence in $L^2[0, T]$ and $d_n \to \infty$ as $n \to \infty$.

Let the coordinates of $h \in L^2[0, T]$ with respect to the basis $\{\psi_r, r \geq 1\}$ be denoted $(h_r, r \geq 1)$ and denote the vector $(h_1, \ldots, h_{d_n})'$ by $\mathbf{h}^{(n)}$. Then, omitting terms not involving h, for $h \in S_n$

$$L^{(n)}(h) = \mathbf{h}^{(n)'}(\mathbf{Q}^{(n)} - \mathbf{P}^{(n)}) - \frac{n}{2}\mathbf{h}^{(n)'}\mathbf{h}^{(n)} \tag{12}$$

where $\mathbf{Q}^{(n)}$ and $\mathbf{P}^{(n)}$ are $d_n \times 1$ vectors with components

$$Q_r^{(n)} = \sum_{i=1}^{n} \int_0^T \psi_r(t)\, dY_i(t)$$

$$P_r^{(n)} = \sum_{i=1}^{n} \int_0^T \psi_r(t) U_i(t)\, dt$$

Maximizing (12) with respect to $h^{(n)}$ we obtain

$$\hat{h}(t) = \sum_{r=1}^{d_n} \hat{h}_r \psi_r(t) \qquad (13)$$

where $\hat{\mathbf{h}}^{(n)} = [\hat{h}_1, \ldots, \hat{h}_{d_n}]'$ is given by

$$\hat{\mathbf{h}}^{(n)} = \frac{1}{n}(\mathbf{Q}^{(n)} - \mathbf{P}^{(n)}) \qquad (14)$$

Theorem 2. Suppose that $d_n \to \infty$ and $d_n/n \to 0$ as $n \to \infty$. Then the orthogonal series sieve estimator $\hat{h}$ given by (13) is a strongly L^2-consistent estimator of h.

Proof: It suffices to show that $\|\hat{\mathbf{h}}^{(n)} - \mathbf{h}^{(n)}\| \xrightarrow{a.s.} 0$, where $\|\cdot\|$ can also denote the euclidean norm, depending on the context. By (10) and (14) the rth component of $\hat{\mathbf{h}}^{(n)} - \mathbf{h}^{(n)}$ is given by

$$(\hat{\mathbf{h}}^{(n)} - \mathbf{h}^{(n)})_r = n^{-1/2} \epsilon_r^{(n)}$$

where

$$\epsilon_r^{(n)} = n^{-1/2} \sum_{i=1}^{n} \int_0^T \psi_r(t)\, d\nu_i(t)$$

for $r = 1, \ldots, d_n$. Thus

$$\|\hat{\mathbf{h}}^{(n)} - \mathbf{h}^{(n)}\|^2 = \frac{1}{n} \sum_{r=1}^{d_n} (\epsilon_r^{(n)})^2$$

Now $\epsilon_r^{(n)}$, $r = 1, \ldots, d_n$ are i.i.d. $N(0, 1)$ r.v.s so that $n\|\hat{\mathbf{h}}^{(n)} - \mathbf{h}^{(n)}\|^2$ has a χ^2 distribution with d_n degrees of freedom. The proof is now completed using the Borel-Cantelli type argument given by Beder (1987, Section 5).

Remark. The rate $d_n = o(n)$ is the best possible for L^2-consistency of the orthogonal series sieve estimators, cf. McKeague (1986) and Beder (1987).

3.2 Kernel Estimators for h

Let K be a bounded kernel function having integral 1, support $[-1, 1]$ and let $b_n > 0$ be a bandwidth parameter. Define

$$\tilde{h}(t) = \frac{1}{b_n} \int_0^T K\left(\frac{t-s}{b_n}\right) d\tilde{H}(s) \tag{15}$$

$$\tilde{H}(t) = \frac{1}{n} \sum_{i=1}^n \left\{ Y_i(t) - \int_0^t U_i(s)\, ds \right\} \tag{16}$$

Here $\tilde{H}(t)$ estimates the function $H(t) = \int_0^t h(s)\, ds$.

Theorem 3. Suppose that $b_n \to 0$ and $b_n n^{1-\delta} \to \infty$ for some $0 < \delta < 1$. Then the kernel estimator $\tilde{h}$ given by (15) is a strongly L^2-consistent estimator of h.

Proof: First note that since h is continuous, $\|h^{(n)} - h\| \to 0$ where $h^{(n)}$ is the following smoothed version of h

$$h^{(n)}(t) = \frac{1}{b_n} \int_0^T K\left(\frac{t-s}{b_n}\right) h(s)\, ds$$

It remains to show that $\|\tilde{h} - h^{(n)}\| \xrightarrow{\text{a.s.}} 0$. From (15)

$$\tilde{h}(t) - h^{(n)}(t) = (nb_n)^{-1/2} \epsilon^{(n)}(t)$$

where

$$\epsilon^{(n)}(t) = \frac{1}{\sqrt{b_n}} \int_0^T K\left(\frac{t-s}{b_n}\right) dW^{(n)}(s)$$

and $W^{(n)} = \sqrt{n}(\tilde{H} - H)$. It follows from (10) and (16) that $W^{(n)}$ is a standard Wiener process for all n. Thus $\epsilon^{(n)}(t)$ is Gaussian with mean zero and variance

$$\frac{1}{b_n} \int_0^T K^2\left(\frac{t-s}{b_n}\right) ds \leq \int_{-1}^1 K^2(u)\, du$$

Fix $\eta > 0$ and let $k > 1/\delta$. Applying Hölder's inequality on $[0, T]$, Fubini's Theorem, and noting that the $2k$-th moment of $\epsilon^{(n)}(t)$ is uniformly bounded in n and t we get

$$E\|\tilde{h} - h^{(n)}\|^{2k} \leq \frac{1}{(nb_n)^k} T^{k-1} \int_0^T E(\epsilon^{(n)}(t))^{2k}\, dt$$

$$= O((nb_n)^{-k})$$

$$= O(n^{-k\delta})$$

By Chebyshev's inequality

$$P(\|\tilde{h} - h^{(n)}\| > \eta) \leq \eta^{-2k} E\|\tilde{h} - h^{(n)}\|^{2k} = O(n^{-k\delta})$$

and since $k\delta > 1$ we have

$$\sum_{n=1}^{\infty} P(\|\tilde{h} - h^{(n)}\| > \eta) < \infty$$

for all $\eta > 0$. The Borel-Cantelli lemma gives $\|\tilde{h} - h^{(n)}\| \xrightarrow{\text{a.s.}} 0$.

3.3   Asymptotic Distribution Results for Estimators of g

Let $\gamma(t, s)$ be the resolvent kernel for $\Gamma(t, s)$, so the unique solution of (8) is given by

$$g(t) = h(t) + \int_0^t \gamma(t, s)h(s)\, ds \qquad (17)$$

see Linz (1985, Theorem 3.3). Note that $h(\cdot)$ may be considered as the output of the linear system

$$z'(t) = [A(t) - C(t)D(t)]z(t) + D(t)g(t) \qquad z(0) = 0$$
$$h(t) = g(t) - C(t)z(t)$$

After a trivial manipulation this may be written as a linear system with input $h(\cdot)$ and output $g(\cdot)$:

$$z'(t) = A(t)z(t) + D(t)h(t) \qquad z(0) = 0$$
$$g(t) = C(t)z(t) + h(t)$$

So we may identify the resolvent kernel γ as

$$\gamma(t, s) = C(t)\Psi_A(t, s)D(s) \qquad (18)$$

where Ψ_A is the transition function of the system $x'(t) = A(t)x(t)$.

Let the estimator of g corresponding to h be denoted $\tilde{g}$, so that $\tilde{g}$ is the solution of the Volterra integral equation

$$\tilde{h}(t) = \tilde{g}(t) + \int_0^t \Gamma(t,s)\tilde{\gamma}(s)\,ds$$

Now we can write $\tilde{g}$ explicitly as

$$\tilde{g}(t) = \tilde{h}(t) + \int_0^t \gamma(t,s)\tilde{h}(s)\,ds \tag{19}$$

Our next result makes use of (19) to derive the asymptotic distribution of γ.

Theorem 4. Suppose that $nb_n \to \infty$ and $nb_n^3 \to 0$. Then for each $0 < t < T$

$$(nb_n)^{1/2}(\tilde{g}(t) - g(t)) \xrightarrow{\mathscr{D}} N(0, \kappa^2)$$

where $\kappa^2 = \int_{-1}^{1} K^2(u)\,du$.

Proof: From (17), (19) and the proof of Theorem 3 we have

$$(nb_n)^{1/2}(\tilde{g}(t) - g(t)) = \epsilon^{(n)}(t) + (nb_n)^{1/2}(h^{(n)}(t) - h(t))$$

$$+ b_n^{1/2}\eta^{(n)}(t) + (nb_n)^{1/2}\int_0^t g(t,s)(h^{(n)}(s) - h(s))\,ds \tag{20}$$

where

$$\eta^{(n)}(t) = \frac{1}{b_n}\int_0^t \int_0^T \gamma(t,s)K\left(\frac{s-v}{b_n}\right)dW^{(n)}(v)\,ds$$

The first term $\epsilon^{(n)}(t)$ on the right-hand side of (20) is Gaussian with mean zero and variance

$$\frac{1}{b_n}\int_0^t K^2\left(\frac{t-s}{b_n}\right)ds \to \kappa^2$$

so that $\epsilon^{(n)}(t) \xrightarrow{\mathscr{D}} N(0, \kappa^2)$. The remaining terms tend to zero in probability. For, using a Fubini theorem for stochastic integrals (see Liptser and Shiryayev, 1977, Theorem 5.15) we have

$$\eta^{(n)}(t) = \frac{1}{b_n}\int_0^T \int_0^t \gamma(t,s)K\left(\frac{s-v}{b_n}\right)ds\,dW^{(n)}(v)$$

so that $\eta^{(n)}(t)$ is Gaussian with mean zero and variance

$$\int_0^T \left\{ \frac{1}{b_n} \int_0^t \gamma(t,s) K\left(\frac{s-v}{b_n}\right) ds \right\}^2 dv \to \int_0^t \gamma^2(t,s)\, ds$$

Thus the third term on the right-hand side of (20) is of order $O_P(\sqrt{b_n})$. Since h is Liptschitz, the second and fourth terms are of order $O(\sqrt{nb_n^3})$.

3.4 An Alternative Estimator for g Based on the Resolvent Equation

An equivalent way of writing equation (17) is

$$g(t) = h(t) + \int_0^t \gamma(t,s)\, dH(s) \qquad (21)$$

so an alternative estimator for g is

$$\tilde{g}^a(t) = \tilde{h}(t) + \int_0^t \gamma(t,s)\, d\tilde{H}(s) \qquad (22)$$

where $\tilde{h}$ and $\tilde{H}$ are given by (15) and (16). Not surprisingly, $\tilde{g}^a(t)$ has the same asymptotic distribution as $\tilde{g}$.

Theorem 5. Suppose that $nb_n \to \infty$ and $nb_n^3 \to \infty$. Then for each $0 < t < T$

$$(nb_n)^{1/2}(\tilde{g}^a(t) - g(t)) \xrightarrow{\mathscr{D}} N(0, \kappa^2)$$

where $\kappa^2 = \int_{-1}^1 K^2(u)\, du$.

Proof: From (21), (22) and the proof of Theorem 3 we have

$$(nb_n)^{1/2}(\tilde{g}^a(t) - g(t)) = \epsilon^{(n)}(t) + (nb_n)^{1/2}(h^{(n)}(t) - h(t))$$
$$+ b_n^{1/2} \int_0^t \gamma(t,s)\, dW^{(n)}(s)$$

By the proof of Theorem 4 the first term $\epsilon^{(n)}(t) \xrightarrow{\mathscr{D}} N(0, \kappa^2)$. The second term is of order $O(\sqrt{nb_n^3})$, as in the proof of Theorem 4. Since the processes $W^{(n)}$, $n \geq 1$ are standard Wiener processes, the last term is of order $O_P(\sqrt{b_n})$.

4 TREND IN THE MEASUREMENT PROCESS — PERIODIC CASE

In this section we consider estimation of g under scheme (II) with $f \equiv 0$ and the functions g, A, B and C assumed to be periodic with period T. We need some preliminary results.

Proposition 6 (Bittanti et al., 1984). If the pair $(A(\cdot), 1)$ is completely controllable and the pair $(A(\cdot), C(\cdot))$ is completely observable then there exists a unique positive T-periodic solution $\bar{P}$ to the Riccati differential equation (4). Moreover, $\bar{\Psi}$, obtained by replacing P by $\bar{P}$ in the definition of Ψ, is uniformly asymptotically stable, that is, there exist positive constants K_1 and K_2 such that

$$|\bar{\Psi}(t, s)| \le K_1 \exp[-K_2(t - s)] \qquad \text{for all } s \le t$$

In the scalar case ($p = q = 1$), the pair $(A(\cdot), 1)$ is always completely controllable, and the pair $(A(\cdot), C(\cdot))$ is completely observable under our assumption that $C(\cdot)$ never vanishes on $[0, T]$, see Rubio (1971, Chapter 5). Thus Proposition 6 can be applied directly in that case. Anyway, the hypotheses of Proposition 6 are very natural in the context of linear systems (see Rubio, 1971, Chapter 5).

We shall need the following assumption:

(A) $\int_{-\infty}^{t} \Psi_A(t, s)\, ds < \infty$ and $\int_{-\infty}^{t} \Psi_A^2(t, s)\, ds < \infty$ for all $t \in [0, T]$

Now introduce the function

$$h_\infty(t) = g(t) + \int_{-\infty}^{t} \bar{\Gamma}(t, s)g(s)\, ds \tag{23}$$

where $\bar{\Gamma}(t, s) = -C(t)\bar{\Psi}(t, s)C(s)\bar{P}(s)$. Also define $\bar{\gamma}(t, s) = C(t)\Psi_A(t, s)C(s)\bar{P}(s)$. The following lemma shows that there is a useful analogy to the important representation (17) in the periodic case, with the functions h_∞ and $\bar{g}$ playing similar roles to h and γ.

Lemma 7
(a) $h_\infty(\cdot)$ is T-periodic.
(b) There exists a positive constant R such that $\sup_{t \in [0,T]} |h_i(t) - h_\infty(t)| = O(e^{-Ri})$ as $i \to \infty$.
(c) If $\int_{-\infty}^{t} \Psi_A(t, s)\, ds < \infty$, then

$$g(t) = h_\infty(t) + \int_{-\infty}^{t} \bar{\gamma}(t, s)h_\infty(s)\, ds \qquad t \in [0, T] \tag{24}$$

Proof: From the standard theory of linear O.D.E.s

$$\bar{\Psi}(t, s) = \exp\left\{\int_s^t [A(u) - C^2(u)\bar{P}(u)]\, du\right\}$$

so that, by the periodicity of A, C and $\bar{P}$, $\bar{\Psi}$ has the property

$$\bar{\Psi}(iT + t, iT + s) = \bar{\Psi}(t, s) \qquad \text{for all } s \le t \tag{25}$$

This property also holds for $\bar{\Gamma}$. Part (a) then follows using the periodicity of g. Next, letting $\bar{h}_i$ be defined by replacing Γ by $\bar{\Gamma}$ in the definition of h_i, we have

$$h_i(t) = g(iT + t) + \int_{-iT}^t \bar{\Gamma}(iT + t, iT + s)g(iT + s)\, ds$$

$$= g(t) + \int_{-iT}^t \bar{\Gamma}(t, s)g(s)\, ds$$

$$= h_\infty(t) - \int_{-\infty}^{-iT} \bar{\Gamma}(t, s)g(s)\, ds \tag{26}$$

Note that $g(s)\bar{\Gamma}(t, s)$ is uniformly asymptotically stable by Proposition 6 and the boundedness of g and C. Thus by (26) and elementary integration, $\sup_{t\in[0,T]}|\bar{h}_i(t) - h_\infty(t)| = O(e^{-Ri})$ as $i \to \infty$. Here and in what follows, R denotes a generic positive constant which does not depend on T and which may change from use to use. To complete the proof of (b) we need to show that $\sup_{t\in[0,T]}|h_i(t) - \bar{h}_i(t)| = O(e^{-Ri})$ as $i \to \infty$. Now,

$$|h_i(t) - \bar{h}_i(t)| \le \int_0^{iT+t} |\Gamma(iT + t, s) - \bar{\Gamma}(iT + t, s)|\, ds$$

$$\le O(1) \int_0^{iT+t} |\bar{\Psi}(iT + t, s) - \Psi(iT + t, s)|\, ds$$

$$+ O(1) \int_0^{iT+t} \bar{\Psi}(iT + t, s)|\bar{P}(s) - P(s)|\, ds \tag{27}$$

since $P(\cdot)$ is bounded by Roitenberg (1974, p. 425). The first term in (27) is bounded above by

$$O(1) \sum_{r=1}^i \int_{(r-1)T}^{rT} \bar{\Psi}(iT + t, s)\left|\exp\left(\int_s^{iT+t} C^2(u)[\bar{P}(u) - P(u)]\, du\right) - 1\right| ds$$

$$+ O(1) \int_{iT}^{iT+t} \bar{\Psi}(iT + t, s)\left|\exp\left(\int_s^{iT+t} C^2(u)[\bar{P}(u) - P(u)]\, du\right) - 1\right| ds$$

Use asymptotic stability of $\ddot{\Psi}$ to bound the sum of the first $[i/2]$ terms above by $O(e^{-iR})$. We can also bound the sum of the remaining terms by $O(e^{-iR})$ as follows. Writing $P_i(u) = P(iT + u)$ for $0 \le u \le T$, use the rate

$$\sup_{u \in [0,T]} |P_i(u) - \bar{P}(u)| = O(e^{-iR}) \tag{28}$$

given by Roitenberg (1974, Theorem 6, p. 431), to obtain for $s \in [(r-1)T, rT]$

$$\int_s^{iT+t} C^2(u)|\bar{P}(u) - P(u)| \, du \le O(1) \sum_{j=r-1}^{i+1} \int_0^T |\bar{P}(u) - P_j(u)| \, du$$

$$\le O(1) \sum_{j=r-1}^{i+1} e^{-Rj} = O(e^{-Rr})$$

uniformly in i. Then, also using the inequality $|e^x - 1| \le 3|x|$ for $|x| \le 1$, the sum of the "remaining terms" above has the form $\sum_{r=[i/2]+1}^{i+1} O(e^{-Rr}) = O(e^{-Ri})$, as required. The last term in (27) is treated in a similar fashion. This proves (b).

Under the hypothesis of part (c), the kernel $\bar{\gamma}$ satisfies $\int_{-\infty}^t \bar{\gamma}(t, s) \, ds < \infty$. Also note that $\bar{\gamma}$ satisfies the property (25). Thus, since g is T-periodic,

$$g(t) = g(iT + t) = h_i(t) + \int_{-iT}^t \gamma(iT + t, iT + s)h_i(s) \, ds$$

$$= h_i(t) + \int_{-iT}^t \bar{\gamma}(t, s)h_i(s) \, ds + O(1) \int_{-iT}^t \Psi_A(t, s)|P_i(s) - \bar{P}(s)| \, ds$$

$$\rightarrow h_\infty(t) + \int_{-\infty}^t \bar{\gamma}(t, s)h_\infty(s) \, ds$$

as $i \rightarrow \infty$, by the dominated convergence theorem, part (b) of the lemma, and (28). This proves (c).

With the help of Lemma 7 it is now possible to develop results analogous to those of Section 3. For the purposes of illustration we shall discuss kernel estimators. Define the kernel estimator $\tilde{h}_\infty$ of h_∞ to be the T-periodic function coinciding with $\tilde{h}$ given by (15). Then, in view of (24), it is natural to estimate g by

$$\tilde{g}(t) = \tilde{h}_\infty(t) + \int_{-\infty}^t \tilde{\gamma}(t, s)\tilde{h}_\infty(s) \, ds, \qquad t \in [0, T] \tag{29}$$

Theorem 8. Suppose that (A) holds. Then the entire statement of Theorem 4 carries over to the periodic case, giving the asymptotic distribution of the estimator $\tilde{g}$ defined by (29).

Proof: The proof is very similar to the scheme (I) case. Use (24) and (29) to obtain a periodic version of (20):

$$(nb_n)^{1/2}(\tilde{g}(t) - g(t)) = \epsilon^{(n)}(t) + (nb_n)^{1/2}(h_n^*(t) - h_\infty(t)) + b_n^{1/2}\eta^{(n)}(t)$$

$$+ (nb_n)^{1/2} \int_{-\infty}^{t} \bar{\gamma}(t, s)(h_n^*(s) - h_\infty(s))\, ds \qquad (30)$$

where

$$\epsilon^{(n)}(t) = \frac{1}{\sqrt{b_n}} \int_0^T K\left(\frac{t-s}{b_n}\right) dW^{(n)}(s)$$

$$W^{(n)} = \sqrt{n}(\tilde{H} - H^{(n)})$$

$$H^{(n)}(t) = \int_0^t \left[\frac{1}{n}\sum_{i=1}^n h_i(s)\right] ds, \qquad t \in [0, T]$$

h_n^* is the T-periodic extension of

$$h_n^*(t) = \frac{1}{b_n} \int_0^T K\left(\frac{t-s}{b_n}\right)\left[\frac{1}{n}\sum_{i=1}^n h_i(s)\right] ds, \qquad t \in [0, T]$$

to the whole real line, and

$$\eta^{(n)}(t) = \frac{1}{b_n} \int_{-\infty}^t \int_0^T \gamma(t, s) K\left(\frac{s-v}{b_n}\right) dW^{(n)}(v)\, ds$$

It follows from (10') and (16) that $W^{(n)}$ is a standard Wiener process for all n, so that, as in the proof of Theorem 4, $\epsilon^{(n)}(t) \xrightarrow{\mathcal{D}} N(0, \kappa^2)$. By Lemma 7(b)

$$\sup_{s \in [0, T]} \left|\frac{1}{n}\sum_{i=1}^n h_i(s) - h_\infty(s)\right| = O(n^{-1})$$

so that, since h_∞ is Lipschitz, the second term on the right-hand side of (30) is of order

$$(nb_n)^{1/2}(h_n^*(t) - h_\infty(t)) = O\left(\frac{b_n}{n}\right)^{1/2} + O(\sqrt{nb_n^3})$$

Using condition (A) it can be shown that the last term on the right-hand side of (30) is of the same order. Using condition (A) again, the third term can be shown to be of order $O_P(\sqrt{b_n})$, as in the proof of Theorem 4.

5 TREND IN THE STATE PROCESS

Throughout this section is is assumed that the trend in the measurement process is zero. We shall introduce an estimator $\tilde{f}$ of the trend f in the state process such that $BIAS(\tilde{f}, 0, t) \to 0$ uniformly in t as $n \to \infty$ a.s. In fact $\tilde{f}$ is shown to be strongly L^2-consistent. We shall only consider the case of observation scheme (I) since our results can be extended easily to scheme (II) along the lines that we extended our results on estimation of g in Section 4.

To estimate f we need to consider (7), which is a linear Volterra integral equation of the first kind. The usual way to deal with such equations is to convert them into Volterra equations of the second kind by differentiation, see Linz (1985, p. 67). In fact, using this technique, we may solve (7) explicitly for f. Since $C(t)$ does not vanish on $[0, T]$, we obtain

$$f(t) = \frac{h'(t)}{C(t)} + F(t)h(t) \tag{31}$$

where

$$F(t) = D(t) - \frac{C'(t)}{C^2(t)} - \frac{A(t)}{C(t)}$$

Thus the problem of estimating f is similar to the problem of estimating g, except that now we need to estimate h' as well as h. We shall only consider kernel estimates of h', although trigonometric series sieve estimators (see Ibragimov and Khasminski, 1980) could equally well be used.

Let K be a kernel function, as in Section 3, but in addition assume that K is differentiable. Let c_n be a bandwidth parameter, different from b_n. Define

$$\tilde{h}'(t) = \frac{1}{c_n^2} \int_0^T K'\left(\frac{t-s}{c_n}\right) d\tilde{H}(s) \tag{32}$$

The following result, stated without proof, is similar to Theorem 3.

Theorem 9. Suppose that $c_n \to 0$ and $c_n n^{1/3-\delta} \to \infty$ where $0 < \delta < \frac{1}{3}$. Then the kernel estimator $\tilde{h}'$ given by (32) is a strongly L^2-consistent estimator of h'.

In view of this result and (31) it is reasonable to estimate f by

$$\tilde{f}(t) = \frac{\tilde{h}'(t)}{C(t)} + F(t)\tilde{h}(t) \tag{33}$$

where $\tilde{h}$, given by (15), is the kernel estimator of h. Under the joint conditions of Theorems 3 and 9 we see that $\tilde{f}$ is a strongly L^2-consistent estimator of f. Finally, we give an asymptotic distribution result for $\tilde{f}$.

Theorem 10. Suppose that $nb_n \to \infty$, $nb_n^3 \to 0$, $nc_n^3 \to 0$, $nc_n^5 \to 0$ and $c_n = o(b_n^{1/3})$. Then for each $0 < t < T$

$$(nc_n^3)^{1/2}(\tilde{f}(t) - f(t)) \xrightarrow{\mathcal{D}} N(0, \sigma^2(t))$$

where

$$\sigma^2(t) = \frac{\int_{-1}^{1} K'(u)^2\, du}{(C(t))^2}$$

Proof: Directly from (31) and (33)

$$(nc_n^3)^{1/2}(\tilde{f}(t) - f(t)) = (C(t))^{-1}(nc_n^3)^{1/2}(\tilde{h}'(t) - h'(t))$$

$$+ F(t)\left(\frac{c_n^3}{b_n}\right)^{1/2} (nb_n)^{1/2}(\tilde{h}(t) - h(t)$$

It can be shown, using a similar approach to the proof of Theorem 4, that the first term on the right-hand side tends in distribution to $N(0, \sigma^2(t))$. Also from the proof of Theorem 4, and using the condition $c_n = o(b_n^{1/3})$, the second term on the right-hand side is seen to be of order $o_P(1)$.

6   THE MULTIVARIATE CASE

In the general case in which the state and measurement processes are p- and q-dimensional, our results are modified in obvious ways to take into account the fact that A, B, C etc. are matrices. The innovations process ν is now a q-dimensional Wiener process and (4) is replaced by the matrix Riccati equation

$$P'(t) = A(t)P(t) + P(t)A(t)^T - P(t)C(t)^T C(t)P(t) + I$$

with initial condition $P(0) = $ covariance matrix of $X(0)$. Here I is the $p \times p$ identity matrix, and "^{T}" denotes "transcript." The Kalman gain is now given by $D(t) = P(t)C(t)^T$.

In Section 3 the q-dimensional version of the orthogonal series sieve estimator $\hat{h}$ is defined (using the same sieve for each component of g) by

$$\hat{h}_k(t) = \sum_{r=1}^{d_n} \hat{h}_{kr} \psi_r(t)$$

$k = 1, \ldots, q$, where

$$\hat{h}_{kr} = \frac{1}{n} \int_0^T \psi_r(t)(dY_{ik}(t) - U_{ik}(t)\,dt)$$

The kernel estimator $\tilde{h}$ is defined (using the same kernel function and bandwidth for each component of g) by the q-dimensional version of (15). The estimators $\hat{g}$ and $\tilde{g}$ are defined as before. Theorems 1–5 extend with the modification that the limiting distribution in Theorems 4 and 5 is $N(0, \kappa^2 I)$. In the proofs of these results, $W^{(n)}$ becomes a q dimensional Wiener process.

For the results of Section 4 to hold, the additional assumptions that $(A(\cdot), I)$ is completely controllable and $(A(\cdot), C(\cdot))$ is completely observable are needed. Condition (A) becomes

(A) $\quad \displaystyle\int_{-\infty}^t \|\Psi_A(t, s)\|\,ds < \infty \quad$ and $\quad \displaystyle\int_{-\infty}^t \|\Psi_A(t, s)\|^2\,ds < \infty$

for all $t \in [0, T]$.

Here $\|\cdot\|$ denotes operator norm. There is essentially no change in the proofs, with the results of Bittanti et al. (1984) and Roitenberg (1974) being applied in the same way as before.

The results of Section 5 extend under the condition that for each $t \in [0, T]$ the matrix $C(t)$ has a left inverse $C^{-1}(t)$. This will be the case if $p \leq q$ and $C(t)$ has column rank p for each $t \in [0, T]$. Then (31) becomes

$$f(t) = C^{-1}(t)h'(t) + F(t)h(t)$$

where

$$F(t) = D(t)C(t)C^{-1}(t) - C^{-1}(t)C'(t)C^{-1}(t) - A(t)C^{-1}(t)$$

showing that f is identifiable. Note that f is not identifiable if $p > q$. The limiting distribution in Theorem 10 becomes $N(0, \Sigma(t))$, where

$$\Sigma(t) = C^{-1}(t)C^{-1}(t)^T \int_{-1}^{1} K'(u)^2 \, du$$

7 DIRECTIONS FOR FURTHER WORK

The techniques and results developed in this article are by no means exhaustive. We are aware of many important questions concerning the problem of nonparametric inference for linear systems in continuous time for which we have no answer at this stage. We conclude by listing some of these questions, the first two of which were mentioned by a referee.

1. Is it possible to weaken the assumption that A, B, and C be known? How far would the analysis go, say, if B was unknown? (This would require an assumption of sufficient variability in the deterministic inputs u_i, $i \geq 1$ to avoid an identifiability problem.) In the same line, how robust are the estimators of f and g to the specification of A, B and C?

2. Can anything be said about the optimal choice of the bandwidth in the kernel estimators? In the cases of density estimation and nonparametric curve estimation there are various techniques for automatically selecting the bandwidth. It ought to be possible to develop such methods of "cross-validation" here.

3. Can a test for detecting the *presence of a trend* (e.g., a test of $g \neq 0$) be developed? More generally, it is of interest to test of whether the trend is of some specified form. As in the case of goodness-of-fit testing for distribution functions, this might be done by deriving a functional central limit theorem for an estimator of the cumulative trend function $G(\cdot) = \int_0^{\cdot} g(s) \, ds$.

REFERENCES

Aihara, S., and Bagchi, A. (1989). Infinite dimensional parameter identification for stochastic parabolic systems. *Statist. Prob. Letters, 8*, 279–287.

Andersen, P. K., Borgan, Ø., Gill, R. D., and Keiding, N. (1988). Censoring, truncation and filtering in statistical models based on counting processes. *Contemporary Math., 80*, 19–59.

Arjas, E., and Haara, P. (1984). A marked point process approach to censored failure data with complicated covariates. *Scand. J. Statist.*, *11*, 193–209.

Bagchi, A., and Borkar, V. (1984). Parameter identification on infinite dimensional linear systems. *Stochastics*, *12*, 201–213.

Balakrishnan, A. V. (1973). *Stochastic Differential Systems*. Lecture notes in economics and mathematical systems. Springer-Verlag, Berlin.

Beder, J. H. (1987). A sieve estimator for the mean of a Gaussian process. *Ann. Statist.*, *15*, 59–78.

Bittanti, S., Colaneri, P., and Guardabassi, G. (1984). Periodic solutions of periodic Riccati equations. *IEEE Trans. Automatic Control AC-29*, 665–667.

Bittanti, S., and Guardabassi, G. (1986). Optimal periodic control and periodic systems analysis: An overview. In *Proc. 25th Conf. Decision Control*, IFAC, Athens, Greece, 1986, pp. 1417–1423.

Davis, M. H. A. (1977). *Linear Estimation and Control*. Chapman and Hall, London.

Davis, M. H. A., and Vinter, R. B. (1985). *Stochastic Modelling and Control*. Chapman and Hall, London.

Geman, S., and Hwang, C. R. (1982). Nonparametric maximum likelihood estimation by the method of sieves. *Ann. Statist.*, *10*, 401–414.

Greenwood, P. E. (1988). Partially specified semimartingale experiments. *Contemporary Math.*, *80*, 1–17.

Grenander, U. (1981). *Abstract Inference*. Wiley, New York.

Ibragimov, I. A., and Khasminski, R. Z. (1980). Estimates of the signal, its derivatives, and point of maximum for Gaussian distributions. *Theory Prob. Appl.*, *25*, 703–719.

Ibragimov, I. A., and Khasminski, R. Z. (1981). *Statistical Estimation*. Springer-Verlag, New York.

Jazwinski, A. H. (1970). *Stochastic processes and filtering theory*. Academic Press, New York.

Kallianpur, G. (1980). *Stochastic Filtering Theory*. Springer-Verlag, New York.

Kalman, R. E., and Bucy, R. S. (1961). New results in linear filtering and prediction theory. *Trans. ASME Ser. D, J. Basic Eng. 83*, 95–108.

Kumar, P. R., and Varaiya, P. (1986). *Stochastic Systems: Estimation, Identification and Adaptive Control*. Prentice-Hall, Englewood Cliffs, N.J.

Leskow, J. (1989). Maximum likelihood estimator for almost periodic stochastic process models. Tech. Report, Center for Stochastic Processes, Dept. of Statistics, University of North Carolina at Chapel Hill.

Linz, P. (1985). *Analytical and Numerical Methods for Volterra Equations.* SIAM, Philadelphia.

Liptser, R. S., and Shiryayev, A. N. (1977). *Statistics of Random Processes I: General Theory.* Springer-Verlag, New York.

Liptser, R. S., and Shiryayev, A. N. (1978). *Statistics of Random Processes: Applications.* Springer-Verlag, New York.

McKeague, I. W. (1986). Estimation for a semimartingale regression model using the method of sieves. *Ann. Statist.*, *14*, 579–589.

McKeague, I. W. (1988). A counting process approach to the regression analysis of grouped survival data. *Stoch. Process. Appl.*, *28*, 221–239.

Nguyen, P. A., and Pham, T. D. (1982). Identification of nonstationary diffusion model by the method of sieves. *SIAM J. Control Optim.*, *20*, 603–611.

Pons, O., and de Turckheim, E. (1988). Cox's periodic regression model. *Ann. Statist.*, 678–693.

Ramlau-Hansen, H. (1983). Smoothing counting process intensities by means of kernel functions. *Ann. Statist.*, *11*, 452–466.

Roitenberg, I. (1974). *Théorie du Controle Automatique.* Mir, Moscow.

Rubio, J. E. (1971). *The Theory of Linear Systems.* Academic Press, New York.

Tugnait, J. K. (1980). Identification and model approximation for continuous-time systems on finite parameter sets. *IEEE Trans. Automatic Control AC-25*, 1202–1206.

6

Weak Convergence of Two-Sided Stochastic Integrals, with an Application to Models for Left Truncated Survival Data

Michael Davidsen
Copenhagen County Hospital, Herlev, Denmark

Martin Jacobsen
University of Copenhagen, Copenhagen, Denmark

A class of additive processes is studied that typically arise as stochastic integrals from s to t when both arguments are allowed to vary. A Skorohod-type topology is introduced on the space of paths for such two-sided integrals and the matching theory of weak convergence is developed and related to usual weak convergence in Skorohod spaces. As an application some asymptotic results, due to M. Woodroofe, for estimators based on left truncated survival data, are reformulated and rederived.

0 INTRODUCTION

The main purpose of the paper is to discuss convergence in distribution of two-sided additive stochastic processes. Such processes arise, for instance, as stochastic integrals from s to t say, when both s and t are allowed to vary. Of course the two-sided integral is completely

This work is based on the thesis written by M. Davidsen for the degree of cand. scient at the University of Copenhagen, with M. Jacobsen as supervisor.

determined by the one-sided integral obtained by keeping one argument fixed and allowing the other to vary. Typically this one-sided process has sample paths belonging to some Skorohod space, right continuous with left limits. However, as will be shown below, the topology on the space of two-sided integral paths inherited from the Skorohod space, will in general depend on the value of the fixed argument. Hence the need for a new approach, to be developed below in the case where $s \leq t$ vary in an open interval so that there is no outstanding candidate, such as an interval endpoint, for a fixed argument value.

In Section 1 we introduce the topology on the space of paths for additive processes. Section 2 treats the matching weak convergence of probabilities on this space. Finally, Section 3 contains an application involving nonparametric estimators of a survival distribution based on left truncated data.

1 THE SPACE $D(\Delta)$

Let $\Delta = \{(s, t): 0 < s \leq t\}$ and let $D(\Delta)$ denote the space of all functions $w: \Delta \to \mathbb{R}$ with the following two properties:

(i) $w(s, u) = w(s, t) + w(t, u)$ $(0 < s \leq t \leq u)$
(ii) $w(s, t)$ is right continuous with left limits in either variable s or t.

Taking $t = u$ in (i) shows that

(iii) $w(t, t) = 0$ $(t > 0)$

It is critical that only *strictly* positive arguments s and t are allowed. The applications we have in mind involve cases, where expressions like $w(0, t)$ are meaningless.

For an obvious example of $D(\Delta)$-functions, let μ be a σ-finite measure on $\mathbb{R}_{++} = (0, \infty)$ and let $f: \mathbb{R}_{++} \to \mathbb{R}$ be a function which is locally μ-integrable in the sense that $\int_K |f| \, d\mu < \infty$ for any compact $K \subset \mathbb{R}_{++}$. Then

$$w(s, t) = \int_{(s,t]} f(u)\mu(du) \tag{1.1}$$

belongs to $D(\Delta)$. Note that since we only consider $s, t > 0$, we allow for $\int_{(0,t]} |f| \, d\mu$ to diverge.

If $w \in D(\Delta)$ it is possible to define $w(s, t)$ also if $s > t > 0$ and still retain properties (i)–(iii), that is, as (1.1) suggests, define

$$w(s, t) = -w(t, s) \tag{1.2}$$

if $0 < t < s$. So from now on we assume $w(s, t)$ to be defined for all $0 < s, t$ and use (i)–(iii) and the antisymmetry (1.2).

Our task in this section is to define a Skorohod-type topology on $D(\Delta)$. First the reader is reminded about the following standard facts and notation.

Let $I \subset \mathbb{R}$ be an arbitrary (bounded or unbounded) interval and let $D(I)$ denote the space of all functions $v: I \to \mathbb{R}$, right continuous with left limits everywhere. The time deformation group for I is the collection Λ_I of bijections $\lambda: I \to I$ which are strictly increasing and continuous. Then $v_n \to v$ in the Skorohod $D(I)$-topology (Skorohod (1956), Billingsley (1968), Lindvall (1973), Whitt (1980) if there exists a sequence $\lambda_n \in \Lambda_I$ such that

(a) $\quad \|\lambda_n - e_I\|_I \to 0$ \hfill (1.3)

(b) $\quad \|v_n \circ \lambda_n - v\|_K \to 0 \qquad$ for all compact intervals $K \subset I$

Here e_I denotes the identity $e_I(t) = t$ on I, and if J is an interval and f is a real-valued function, defined on a domain containing J, we write

$$\|f\|_J = \sup_{t \in J} |f(t)|$$

If $I = \mathbb{R}_{++}$, we omit index I. In particular Λ is the time deformation group on $\mathbb{R}_{++}$. Also for $I = [t_0, \infty)$, we write index t_0 instead of $[t_0, \infty)$.

Recall that (1.3a) may be replaced by

$$\|\lambda_n - e_I\|_K \to 0 \qquad \text{for all compact intervals } K \subset I \tag{1.3a'}$$

Recall also that $D(I)$ with the Skorohod topology is metrizable as a Polish space.

Returning now to the space $D(\Lambda)$, fix $s_0 > 0$ and consider the map $\psi_{s_0}: D(\Delta) \to D(\mathbb{R}_{++})$ given by

$$(\psi_{s_0} w)(t) = w(s_0, t) \qquad (t \in \mathbb{R}_{++})$$

It is immediately checked that ψ_{s_0} is a bijection from $D(\Delta)$ onto $D_{s_0}(\mathbb{R}_{++}) = \{v \in D(\mathbb{R}_{++}): v(s_0) = 0\}$ with inverse

$$(\psi_{s_0}^{-1} v)(s, t) = v(t) - v(s) \qquad (s, t \in \mathbb{R}_{++})$$

At first glance it would appear natural to equip $D(\Delta)$ with the topology obtained by using on $D_{s_0}(\mathbb{R}_{++})$ the topology it inherits as a subspace of the Skorohod $D(\mathbb{R}_{++})$-space, and then demanding that ψ_{s_0} be a homeomorphism. A quick check reveals however, that this topology depends on the choice of s_0. We therefore take a different approach and present the following basic definition.

1.4 Definition. A sequence $(w_n)_{n \geq 1}$ of $D(\Delta)$-functions converges to $w \in D(\Delta)$ in the Skorohod-$D(\Delta)$ topology if and only if there exists a sequence (λ_n) of time deformations in Λ such that

(a) $\|\lambda_n \rightarrow e\| \rightarrow 0$

(b) $\displaystyle\sup_{s \in K, t \in L} |w_n(\lambda_n s, \lambda_n t) - w(s, t)| \rightarrow 0$ (1.5)

for all compact intervals $K, L \subset \mathbb{R}_{++}$

Various equivalent forms are available: in (b) it suffices to take $K = L$ and (a) may be replaced by (1.3a') (with $I = \mathbb{R}_{++}$).

For convenience we shall write $\|w_n \circ \lambda_n - \varphi\|_{K,L}$ for the supremum in (1.5b).

It should be clear that Definition 1.4 defines a topology on $D(\Delta)$: indeed, a subbase of neighborhoods of $w \in D(\Delta)$ is given by the collection of sets of the form

$$U(\epsilon, K, L, w) = \{w' \in D(\Delta) : \exists \lambda \in \Lambda \qquad \text{such that}$$

$$\|\lambda - e\| < \epsilon, \|w' \circ \lambda - w\|_{K,L} < \epsilon\}$$

for arbitrary $\epsilon > 0$ and $K, L \subset \mathbb{R}_{++}$ compact intervals.

Note that s, t are treated symmetrically in (1.5b). This is not the case with the topology on $D(\Delta)$ described above, which was derived from the Skorohod topology on $D_{s_0}(\mathbb{R}_{++})$.

Before listing some properties of the Skorohod $D(\Delta)$-topology, we need to discuss the discontinuities of $D(\Delta)$-functions.

Let $w \in D(\Delta)$. For $s_0 > 0$ fixed, $\psi_{s_0} w$ has at most countably many points of discontinuity. Furthermore, since

$$(\psi_{s_0} w)(t) - (\psi_{t_0} w)(t) = w(s_0, t_0)$$

the discontinuity set does not depend on s_0, and we are allowed to define the set of *continuity points* for $w \in D(\Delta)$ as

$$C(w) = \{t \in \mathbb{R}_{++} : \psi_{s_0} w \text{ is continuous at } t\}$$ (1.6)

for any $s_0 > 0$.

It is immediate that $w \in D(\Delta)$ is continuous at $(s, t) \in \Delta$ iff $s, t \in C(w)$.

1.7 Proposition. (a) The following three conditions are equivalent:

(i) $w_n \to w$ in $D(\Delta)$;

(ii) for some, and then automatically for all $t_0 > 0$, there exists a sequence of Λ-functions such that

$$\|\lambda_n - e\| \to 0,$$

$$\|(\psi_{\lambda_n t_0} w_n) \circ \lambda_n - \psi_{t_0} w\|_L \to 0 \text{ for all compact} \tag{1.8}$$

intervals $L \subset \mathbb{R}_{++}$;

(iii) for some, and then automatically for all $t_0 \in C(w)$, $\psi_{t_0} w_n \to \psi_{t_0} w$ in $D(\mathbb{R}_{++})$.

(b) For any $t_0 > 0$, the mapping $\psi_{t_0}: D(\Delta) \to D_{t_0}(\mathbb{R}_{++})$ is continuous at $w \in D(\Delta)$ provided $t_0 \in C(w)$. The inverse mapping $\psi_{t_0}^{-1}: D_{t_0}(\mathbb{R}_{++}) \to D(\Delta)$ is everywhere continuous.

Proof. (a) Taking $K = \{t_0\}$ in (1.5b) shows that (i) $\Rightarrow$ (ii) for all t_0. Conversely, if (ii) holds for some $t_0 > 0$, find (λ_n) such that (1.8) is true. Since by additivity and antisymmetry

$$|w_n(\lambda_n s, \lambda_n t) - w(s, t)|$$
$$\leq |w_n(\lambda_n t_0, \lambda_n s) - w(t_0, s)| + |w_n(\lambda_n t_0, \lambda_n t) - w(t_0, t)|$$

(1.5b) and (i) follows.

Assume now that (ii) holds, and take an arbitrary $t_0 \in C(w)$. Find (λ_n) such that (1.8) holds for this t_0. To establish (iii) we must show that for all L compact,

$$\|(\psi_{t_0} w_n) \circ \lambda_n - \psi_{t_0} w\|_L \to 0 \tag{1.9}$$

But for all t,

$$((\psi_{t_0} w_n) \circ \lambda_n)(t) - (\psi_{t_0} w)(t)$$
$$= (\psi_{\lambda_n t_0} w_n(\lambda_n t) - (\psi_{t_0} w)(t)) + w_n(t_0, \lambda_n t_0)$$

so it suffices to show that $w_n(t_0, \lambda_n t_0) \to 0$. But since $\lambda_n^{-1} t_0 \to t_0$, using (1.8) and the assumption that $t_0 \in C(w)$, it follows easily that $w_n(t_0, \lambda_n t_0) \to w(t_0, t_0) = 0$.

That (iii) for one $t_0 \in C(w)$ implies (i), follows trivially from the last assertion in (b) proved below. The proof of part (a) is complete.

(b) The first assertion is just the implication (i) $\Rightarrow$ (iii). For the second, let $t_0 > 0$ and assume v_n, v are $D_{t_0}(\mathbb{R}_{++})$-functions such that

$v_n \to v$ in $D(\mathbb{R}_{++})$. Find $\lambda_n \in \Lambda$ such that $\|\lambda_n - e\| \to 0$ and $\|v_n \circ \lambda_n - v\|_K \to 0$ for all compact $K \subset \mathbb{R}_{++}$. Then for $K, L \subset \mathbb{R}_{++}$ compact

$$\|(\psi_{t_0}^{-1} v_n) \circ \lambda_n - \psi_{t_0}^{-1} v\|_{K,L}$$
$$= \sup_{s \in K, t \in L} |(v_n(\lambda_n t) - v_n(\lambda_n s)) - (v(t) - v(s))| \to 0$$

From the proposition it follows in particular that if $w_n \to w$, then $w_n(s, t) \to w(s, t)$ if $s, t \in C(w)$.

1.10 Proposition. The Skorohod space $D(\Delta)$ is metrizable as a Polish space.

We shall only outline the proof. For $0 < s < t$ define a distance between the restrictions to $[s, t]$ of two $D(\Delta)$-functions w_1, w_2 by

$$d_{st}(w_1, w_2) = \inf_{\lambda \in \Lambda_{st}} (\|\lambda - e\|_{st} \vee \|w_1 - w_2 \circ \lambda\|_{st})$$

where st is short for $[s, t]$ and

$$\|w_1 - w_2 \circ \lambda\|_{st} = \sup_{u, u' \in [s,t]} |w_1(u, u') - w_2(\lambda u, \lambda u')|$$

Then (see Whitt (1980), Section 2),

$$d(w_1, w_2) = \int_0^1 \int_1^\infty e^{s-t} (d_{st}(w_1, w_2) \wedge 1) \, dt \, ds$$

defines a metric for the Skorohod-$D(\Delta)$-topology. Thus $D(\Delta)$ is metrizable and it is then readily seen, that it is separable: convert a countable, dense subset A of $D(\mathbb{R}_{++})$ to a countable, dense subset A' of $D_{t_0}(\mathbb{R}_{++})$ by subtracting from any $v \in A$ the constant $v(t_0)$, and note that since $\psi_{t_0}^{-1}$ is continuous (Proposition 1.7(b)) and onto, $\psi_{t_0}^{-1}(A')$ is countable and dense in $D(\Delta)$.

Finally, to prove completeness, one must modify the metric d by considering only time deformations that are not too steep, exactly as in Billingsley (1968), p. 113.

2 WEAK CONVERGENCE

Let P_n for $n \geq 1$ and P be probabilities on a metric space. We write $P_n \Rightarrow P$ if P_n converges weakly to P as $n \to \infty$, that is

$$\int f \, dP_n \rightarrow \int f \, dP$$

for all bounded and continuous f.

We shall now discuss weak convergence of probabilities on $D(\Delta)$. Since $D(\Delta)$ is Polish, Prohorov's theorem applies and consequently $P_n \Rightarrow P$ iff there is weak convergence of all finite-dimensional distributions corresponding to continuity points for P (see below) and the family (P_n) is tight.

We shall not here discuss conditions for tightness, but instead relate weak convergence of probabilities on $D(\Delta)$ to the standard case of weak convergence of probabilities on $D(\mathbb{R}_{++})$ (or rather $D[t_0, \infty)$ for almost all $t_0 > 0$).

Let $t > 0$, let r_t denote the map that restricts a function with a domain containing $[t, \infty)$ to $[t, \infty)$, for example, $r_t v = (v(u))_{u \geq t}$ for $v \in D(\mathbb{R}_{++})$. Also, write $\eta_t = r_t \circ \psi_t$, so that $\eta_t \colon D(\Delta) \rightarrow D[t, \infty)$ and

$$(\eta_t w)(u) = w(t, u) \qquad (u \geq t) \tag{2.1}$$

If P is a probability on $D(\Delta)$, introduce T_P as the points of continuity for P, that is

$$T_P = \{t > 0 \colon P(C_t) = 1\}$$

where

$$C_t = \{w \in D(\Delta) \colon t \in C(w)\}$$

Recalling that $t \in C(w)$ iff t is a continuity point for one (and then all) $\psi_{s_0} w$ (see (1.6)), and using standard properties of probabilities on $D(\mathbb{R}_{++})$, it follows that T_P is dense in $\mathbb{R}_{++}$.

The main result we shall prove is the following.

2.2 Theorem. For $(P_n)_{n \geq 1}$, P probabilities on $D(\Delta)$, $P_n \Rightarrow P$ if and only if $\eta_t(P_n) \Rightarrow \eta_t(P)$ for all $t \in T_P$.

Notation. $\eta_t(P_n)$, $\eta_t(P)$ are P_n, P transformed by η_t.

Proof. Whitt (1980), Theorem 2.8, showed that weak convergence on $D(\mathbb{R}_{++})$ amounts (essentially) to weak convergence on $D[s, t]$ for all $0 < s < t$. We follow his proof in order to obtain the nontrivial half of the theorem.

If $P_n \Rightarrow P$, trivially $\eta_t(P_n) \Rightarrow \eta_t(P)$ for $t \in T_P$ because $\eta_t = r_t \circ \psi_t$ is P-

a.s. continuous, since ψ_t is a.s. continuous by Proposition 1.7(b), and r_t is continuous at $v \in D(\mathbb{R}_{++})$, whenever v is continuous at t.

Suppose conversely that $\eta_t(P_n) \Rightarrow \eta_t(P)$ for all $t \in T_P$. In order that $P_n \Rightarrow P$ it is necessary and sufficient that

$$\limsup_{n \to \infty} P_n(F) \le P(F) \tag{2.3}$$

for any closed set $F \subset D(\Delta)$.

For $t > 0$, write $H_t = \eta_t^{-1}(\overline{\eta_t F})$ where $\overline{\eta_t F}$ is the closure in $D[t, \infty)$ of the image of F under η_t. Since T_P is dense, we can choose a sequence $t_k \downarrow 0$ of points in T_P. Write $H_k = H_{t_k}$. Now $H_k \supset F$ and since $\eta_{t_k}(P_n) \Rightarrow \eta_{t_k}(P)$ as $n \to \infty$, for all k

$$\limsup_{n \to \infty} P_n(F) \le \limsup_{n \to \infty} P_n(H_k) = \limsup_{n \to \infty} \eta_{t_k}(P_n)(\overline{\eta_{t_k} F})$$

$$\le \eta_{t_k}(P)(\overline{\eta_{t_k} F}) = P(H_k)$$

Thus (2.3) will follow, if we show that

$$H_k \supset H_{k+1} \; P\text{-a.s. for all } k \tag{2.4}$$

$$\bigcap_{k=1}^{\infty} H_k \subset F \tag{2.5}$$

For (2.4), because $t_k, t_{k+1} \in T_P$, it is enough to show that if $w \in H_{k+1}$ is such that $t_k, t_{k+1} \in C(w)$, then $w \in H_k$. First note that for any $w' \in D(\Delta)$,

$$\eta_{t_k} w' = r_{t_k}(\eta_{t_{k+1}} w') - w'(t_{k+1}, t_k) \tag{2.6}$$

Since $\eta_{t_{k+1}} w \in \overline{\eta_{t_{k+1}} F}$ we can find a sequence (w_n) from F such that $\eta_{t_{k+1}} w_n \to \eta_{t_{k+1}} w$ in $D[t_{k+1}, \infty)$ as $n \to \infty$. Because $t_k \in C(w)$, in particular $\eta_{t_{k+1}} w$ is continuous at t_k, and therefore r_{t_k} is continuous at $\eta_{t_{k+1}} w$. From (2.6) it follows that as $n \to \infty$,

$$\eta_{t_k} w_n \to r_{t_k}(\eta_{t_{k+1}} w) - w(t_{k+1}, t_k) = \eta_{t_k} w$$

that is, $\eta_{t_k} w \in \overline{\eta_{t_k} F}$, which is exactly to say that $w \in H_k$.

To show (2.5), assume that $w \in \cap H_k$. We shall exhibit a sequence (w_k) from F such that $w_k \to w$. Since F is closed, $w \in F$ then follows. By assumption, for any k, $\eta_{t_k} w \in \overline{\eta_{t_k} F}$, and we can find a sequence $(w_{kn})_{n \ge 1}$ from F such that $\eta_{t_k} w_{kn} \to \eta_{t_k} w$ as $n \to \infty$. By the definition of convergence in $D[t_k, \infty)$, this implies that there exists integers n_k and $\lambda'_k \in \Lambda_{t_k}$ such that

$$\|\lambda'_k - e_{t_k}\|_{t_k} \le \frac{1}{k} \tag{2.7}$$

$$\sup_{t \in [t_k, k]} |w_{kn_k}(t_k, \lambda'_k t) - w(t_k, t)| \le \frac{1}{k} \tag{2.8}$$

Using Proposition 1.7(a), we show that $w_{kn_k} \to w$ in $D(\Delta)$ by showing that $\psi_{t_1} w_{kn_k} \to \psi_{t_1} w$ in $D(\mathbb{R}_{++})$ as $k \to \infty$. To this end, define for $k = 1, 2, \dots$.

$$\lambda_k t = \begin{cases} \lambda'_k t & t \ge t_k \\ t & 0 < t \le t_k \end{cases}$$

Clearly, $\lambda_k \in \Lambda$ and $\|\lambda_k - e\| = \|\lambda'_k - e_{t_k}\|_{t_k} \to 0$ by (2.7). It remains to show uniform convergence on compacts of $\psi_{t_1} w_{kn_k}$ to $\psi_{t_1} w$. Since any compact set is contained in $[t_k, k]$ for k large enough, for this it clearly suffices to show that

$$\sup_{t \in [t_k, k]} |w_{kn_k}(t_1, \lambda_k t) - w(t_1, t)| \to 0$$

But the supremum is

$$\le \sup_{t \in [t_k, k]} |w_{kn_k}(t_k, \lambda_k t) - w(t_k, t)| + |w_{kn_k}(t_k, t_1) - w(t_k, t_1)|$$

The first term is $\le 1/k$ by (2.8). The second equals

$$|w_{kn_k}(t_k, \lambda_k(\lambda_k^{-1} t_1)) - w(t_k, t_1)|$$

$$\le |w_{kn_k}(t_k, \lambda_k(\lambda_k^{-1} t_1)) - w(t_k, \lambda_k^{-1} t_1)| + |w(\lambda_k^{-1} t_1, t_1)|$$

Because of (2.7), $\lambda_k^{-1} t_1 \to t_1$, so by (2.8) the first term is $\le 1/k$ for k large. The second $\to 0$ because $t_1 \in C(w)$.

Suppose now that $(X_n)_{n \ge 1}$, X are real valued stochastic processes, $X_n = (X_n(s, t))_{s,t > 0}$, $X = (X(s, t))_{s,t > 0}$ with sample paths in $D(\Delta)$. Write P_n, P for the distribution of X_n, X respectively and $X_n \overset{d}{\to} X$ if X_n converges in distribution to X, that is, if $P_n \Rightarrow P$. With $T_X = T_P$ the theorem may be restated as follows: $X_n \overset{d}{\to} X$ (in $D(\Delta)$) iff $\eta_{t_0} X_n \overset{d}{\to} \eta_{t_0} X$ (in $D[t_0, \infty)$) for every $t_0 \in T_X$.

We shall conclude this section with a discussion of when it is possible to include $t_0 = 0$ so as to deduce weak convergence on $D[0, \infty)$ from weak convergence on $D(\Delta)$.

For this problem to make sense at all, it is necessary that the X_n and X in a natural fashion extend to processes defined also at time 0. More specifically, assume that for some (and then as seen below, automatically for all) $t_0 > 0$, the limits

$$X_{+n}(t_0) = \lim_{h \downarrow 0} X_n(h, t_0) \qquad (n \geq 1) \qquad (2.9)$$

$$X_{+}(t_0) = \lim_{h \downarrow 0} X(h, t_0) \qquad\qquad (2.10)$$

exist almost surely.

Using additivity it follows immediately that if (2.9) holds, then a.s. $X_{+n}(t) = \lim_{h \downarrow 0} X_n(h, t)$ exists simultaneously for all t. Also

$$X_{+n}(t) - X_{+n}(s) = X_n(s, t),$$

$$\lim_{h \downarrow 0} X_{+n}(h) = 0 \text{ a.s.}$$

Thus we may define processes $X_{+n} = (X_{+n}(t))_{t \geq 0}$, $X_{+} = (X_{+}(t))_{t \geq 0}$ with paths in $D[0, \infty)$, always taking the value 0 for $t = 0$, and may then ask when $X_{+n} \xrightarrow{d} X_{+}$ (in $D[0, \infty)$), assuming that $X_n \xrightarrow{d} X$.

First, consider the following simple example: P_n, P are degenerate with unit mass at $w_n, w \in D(\Delta)$ respectively, where

$$w_n(s, t) = \int_s^t a_n(1_{[1/n, 2/n]}(u) - 1_{[2/n, 3/n]}(u)) \, du \qquad w \equiv 0$$

No matter what are the constants a_n, $w_n \to w$ in $D(\Delta)$ and thus $P_n \Rightarrow P$. Also, (2.8), (2.9) hold and almost surely

$$X_{+n}(t) = \begin{cases} 0 & \left(t \leq \dfrac{1}{n}\right) \\[2mm] a_n\left(t - \dfrac{1}{n}\right) & \left(\dfrac{1}{n} < t \leq \dfrac{2}{n}\right) \\[2mm] a_n\left(\dfrac{3}{n} - t\right) & \left(\dfrac{2}{n} < t \leq \dfrac{3}{n}\right) \\[2mm] 0 & \left(t > \dfrac{2}{n}\right) \end{cases} \qquad X_{+}(t) = 0$$

From this it is clear that the finite-dimensional distributions of X_{+n}

converge to those of X_+, but it is also clear that X_{+n} need not converge in distribution to X_+, indeed $X_{+n} \xrightarrow{d} X_+$ iff $a_n/n \to 0$.

Thus, to obtain the desired $D[0, \infty)$-convergence, one must be able to control all process values close to 0. The precise formulation as given in condition (2.12) below, may be referred to as "tightness close to 0."

2.11 Theorem Let $(X_n)_{n \geq 1}$, X be processes with sample paths in $D(\Delta)$ such that $X_n \xrightarrow{d} X$ and (2.9), (2.10) hold almost surely. In order that $X_{+n} \xrightarrow{d} X_+$ (in $D[0, \infty)$) it is necessary and sufficient that the following condition holds:

$$\forall \epsilon, \ \eta > 0 \exists \delta > 0, \ n_0 \in \mathbb{N} \quad \forall n \geq n_0$$

$$\Pr(\sup_{0 < s \leq t \leq \delta} |X_n(s, t)| > \epsilon) < \eta \tag{2.12}$$

Notation. The processes X_n, X may be defined on different probability spaces, but we always write Pr for the relevant probability and E for the relevant expectation.

Proof. Assume first that $X_{+n} \xrightarrow{d} X_+$ (in $D[0, \infty)$). In particular, for a given $0 < t_0 \in T_X$, there is weak convergence (in $D[0, t_0]$) of the restrictions of the X_{+n} to $[0, t_0]$, towards the restriction of X_+. In particular, by (15.8) in Billingsley (1968), Theorem 15.3,

$$\forall \epsilon, \ \eta > 0 \quad \exists 0 < \delta < t_0, \ n_0 \in \mathbb{N} \ \forall n \geq n_0$$

$$\Pr(\sup_{0 \leq s \leq t \leq \delta} |X_{+n}(t) - X_{+n}(s)| > \epsilon) < \eta$$

Since $X_{+n}(t) - X_{+n}(s) = X_n(s, t)$, this is precisely (2.12).

Suppose now that we have tightness close to 0. In order to show $X_{+n} \xrightarrow{d} X_+$, we must show that for all $t_0 \in T_X$, $(X_{+n}(t))_{0 \leq t \leq t_0}$ $\xrightarrow{d} (X_+(t))_{0 \leq t \leq t_0}$ (in $D[0, t_0]$). For this, by Billingsley (1968), Theorem 15.4, it is enough to show that (i) the finite-dimensional distributions of X_{+n} converge to those of X_+ when all time points involved belong to T_X; and (ii) the following condition holds for any $t_0 \in T_X$:

$$\forall \epsilon, \ \eta > 0 \exists 0 < \delta < t_0, \ n_0 \in \mathbb{N} \ \forall n \geq n_0 \tag{2.13}$$

$$\Pr(\sup_{\substack{0 \leq t_1 \leq t \leq t_2 \leq t_0 \\ |t_2 - t_1| \leq \delta}} |X_{+n}(t) - X_{+n}(t_1)| \wedge |X_{+n}(t_2) - X_{+n}(t)| > \epsilon) < \eta$$

Proof of (i). Because $X_{+n}(0)$, $X_+(0)$ both $= 0$ a.s., we need only show that for any choice of k and $0 < t_1 \leq \cdots \leq t_k \in T_X$,

$$(X_{+n}(t_1), \ldots, X_{+n}(t_k)) \xrightarrow{d} (X_+(t_1), \ldots, X_+(t_k)) \qquad (2.14)$$

Given $\epsilon, \eta > 0$ find $0 < \delta < t_1$, $\delta \in T_X$ such that (2.12) holds for n sufficiently large. Because $X_n \xrightarrow{d} X$,

$$(X_n(\delta, t_1), \ldots, X_n(\delta, t_k)) \xrightarrow{d} (X(\delta, t_1), \ldots, X(\delta, t_k)) \qquad (2.15)$$

By (2.12) the random vector on the left-hand side of (2.14) is close to the vector on the left of (2.15) with high probability. Because $X_+(t) = \lim_{\delta \downarrow 0} X(\delta, t)$, the same is true for the two right-hand sides if δ is small, and (2.14) follows easily.

Proof of (ii). Since $X_n \xrightarrow{d} X$, by Theorem 2.2, given $0 < s_0 < t_0$, s_0, $t_0 \in T_X$, $(X_n(s_0, t))_{s_0 \leq t \leq t_0} \xrightarrow{d} (X(s_0, t))_{s_0 \leq t \leq t_0}$ (in $D[s_0, t_0]$), so by (15.7) in Billingsley (1968), Theorem 15.3.

$$\forall \epsilon, \; \eta > 0 \; \exists 0 < \delta < t_0 - s_0, \; n_0 \in \mathbb{N} \; \forall n \geq n_0 \qquad (2.16)$$

$$\Pr\left(\sup_{\substack{s_0 \leq t_1 \leq t \leq t_2 \leq t_0 \\ t_2 - t_1 \leq \delta}} |X_n(s_0, t) - X_n(s_0, t_1)| \wedge |X_n(s_0, t_2) - X_n(s_0, t)| > \frac{\epsilon}{2} \right) < \frac{\eta}{2}$$

Note that since, for example, $X_{+n}(t) - X_{+n}(t_1) = X_n(s_0, t) - X_n(s_0, t_1) = X_n(t_1, t)$, (2.12) formally emerges from (2.16) by replacing s_0 by 0.

Write S_n for the supremum appearing in (2.16) and introduce

$$R_n = \sup_{0 < s \leq t \leq s_0} |X_n(s, t)|$$

Then whenever $0 \leq t_1 \leq t \leq t_2 \leq t_0$, $t_2 - t_1 \leq \delta$ the minimum of differences in (2.13) equals

$$|X_n(t_1, t)| \wedge |X_n(t, t_2)| \leq \begin{cases} R_n & \text{if } t \leq s_0 \\ R_n + S_n & \text{if } t_1 \leq s_0 < t \\ S_n & \text{if } s_0 < t_1 \end{cases} \qquad (2.17)$$

and therefore is always $\leq R_n + S_n$. Thus to prove (2.13) for $t_0 \in T_X$, given $\epsilon, \eta > 0$ first use (2.12) to choose $0 < s_0 < t_0$, $s_0 \in T_X$, so that $\Pr(R_n > \epsilon/2) < \eta/2$ for n sufficiently large. Then pick δ in accordance with (2.16) and use (2.17) to arrive at (2.13).

3 AN APPLICATION TO MODELS FOR LEFT TRUNCATED SURVIVAL DATA

In this section we review some results due to Woodroofe (1985), using the theory of the preceding sections.

Let F, G denote two distribution functions for probabilities on $\mathbb{R}_{++}$ and consider n i.i.d. pairs of $\mathbb{R}_{++}$-valued random variables $(X_1, Y_1), \ldots, (X_n, Y_n)$, where the distribution of (X_i, Y_i) is that of a pair (U, V) conditionally on the event $(V < U)$ with U, V independent and having distribution functions F and G respectively. We assume of course that

$$\alpha = \Pr(V < U) > 0$$

Thus in particular $Y_i < X_i$ a.s.

Referring to each i as an item, X_i is the failure time and Y_i the truncation time for item i. We say that i is at risk at time $t > 0$ if $Y_i \leq t < X_i$, corresponding to $I_i(t) = 1$ where

$$I_i(t) = 1_{(Y_i \leq t < X_i)}$$

Now consider the process N_n counting the number of observed failures,

$$N_n(t) = \sum_{i=1}^{n} 1_{(X_i \leq t)}$$

and denote by $\mathscr{F}_t$ the σ-algebra generated by all observations before t, that is, $(N_n(s))_{s \leq t}$ and the events $(Y_i \leq s)$ for $i = 1, \ldots, n$, $s \leq t$.

Assume from now on that F is absolutely continuous with hazard function μ, so that

$$F(t) = 1 - \exp\left(- \int_0^t \mu(s)\,ds \right)$$

Also, for convenience assume that $\int_0^t \mu < \infty$ and that $G(t) > 0$ for all $t \in \mathbb{R}_{++}$.

It is standard [e.g., Keiding and Gill (1990)], that with respect to the filtration $(\mathscr{F}_t)_{t \geq 0}$, the increasing process N_n has compensator

$$\Lambda_n(t) = \int_0^t \mu(s) 1_{(R_n(s) > 0)}\,ds$$

where $R_n(s) = \sum_{i=1}^{n} I_i(s)$. Also then, the integrated hazard $\int_0^t \mu$ may be estimated by the Nelson-Aalen estimator

$$\hat{\beta}_n(t) = \int_{(0,t]} \frac{1}{R_n(s-)} N_n(ds)$$

and in particular, defining

$$\beta_n(t) = \int_0^t \mu(s) 1_{(R_n(s)>0)} \, ds$$

$\hat{\beta}_n - \beta_n$ is a martingale.

Now introduce the process X_n,

$$X_n(s, t) = \sqrt{n}((\hat{\beta}_n(t) - \beta_n(t)) - (\hat{\beta}_n(s) - \beta_n(s)))$$

$$= \sqrt{n} \left(\int_{(s,t]} \frac{1}{R_n(u-)} N_n(du) - \int_s^t \mu(u) 1_{(R_n(u)>0)} \, du \right)$$

with paths in $D(\Delta)$. Woodroofe (1985), Theorem 3 showed that for any $t_0 > 0$,

$$\eta_{t_0} \circ X_n \xrightarrow{d} X_{t_0} \qquad (3.1)$$

in $D[t_0, \infty)$, where $X_{t_0} = (X_{t_0}(t))_{t \geq t_0}$ is the mean zero Gaussian process with independent increments and variance function $EX_{t_0}^2(t) = \sigma^2(t_0, t)$, where

$$\sigma^2(t_0, t) = \alpha \int_{t_0}^t \frac{1}{G(u)(1 - F(u))} \mu(u) \, du$$

Recall that

$$\alpha = \Pr(V < U) = \int_0^\infty G(u)(1 - F(u))\mu(u) \, du$$

In (3.1) the asymptotics are given for the stochastic integrals $\hat{\beta}_n - \beta_n$ from t_0 and out. By Theorem 2.2, Woodroofe's result may immediately be restated as follows:

$$X_n \xrightarrow{d} X \qquad (3.2)$$

with $X = (X(s, t))_{0 < s, t}$ the continuous mean zero Gaussian process, uniquely determined by the requirements that it has paths in $D(\Delta)$, that $X(s, t), X(u, v)$ are independent whenever $s \leq t \leq u \leq v$ and that the variance function is

$$EX^2(s, t) = \sigma^2(s, t) \quad (s, t) \in \Delta$$

(3.1) is proved easily via a suitable functional martingale central limit theorem. In particular, for $t_0 > 0$ the martingale $\eta_{t_0} \circ X_n$ on $[t_0, \infty)$ has quadratic characteristic

$$\langle \eta_{t_0} \circ X_n \rangle(t) = \int_{t_0}^t \frac{n}{R_n(u)} 1_{(R_n(u)>0)} \mu(u) \, du \qquad (3.3)$$

converging in probability to $\sigma^2(t_0, t)$ as $n \to \infty$. To argue this, note that by the law of large numbers, the integrand in (3.3) converges pointwise to

$$\mu(u)/\Pr(V \leq u < U \mid V < U) = \alpha\mu(u)(G(u)(1 - F(u))^{-1}$$

Gill's (1984) useful concept of convergence, boundedly in probability, now allows us to change the order of taking limits in probability and integrating.

Assuming that

$$\int_0^\infty \frac{1}{G} \, dF < \infty \qquad (3.4)$$

Woodroofe (1985), Theorem 5 showed that (3.1) holds, even for $t_0 = 0$. Here we shall derive the same result by verifying the condition for tightness close to 0 from Theorem 2.11.

Clearly (2.9) is satisfied for all n, while (2.10) holds iff $\lim_{s \downarrow 0} \sigma^2(s, t)$ exists and is finite for some $t > 0$. It is immediately verified that this condition is equivalent to (3.4). So assume that (3.4) holds and let $\epsilon, \eta > 0$ be given. Our aim is to find δ, n_0 such that (2.12) holds for $n \geq n_0$. But

$$\Pr(\sup_{s,t:0<s\leq t\leq\delta} |X_n(s, t)| > \epsilon) = \lim_{s \downarrow 0} \Pr(\sup_{t:s\leq t\leq\delta} |X_n(s, t)| > \epsilon) \quad (3.5)$$

and by Doob's inequality applied to the martingale $\eta_s \circ X_n$, the probability on the right is dominated by

$$\frac{1}{\epsilon^2} E(\sup_{t:s\leq t\leq\delta} X_n^2(s, t)) \leq \frac{4}{\epsilon^2} EX_n^2(s, \delta)$$

$$= \frac{4}{\epsilon^2} E\langle \eta_s \circ X_n \rangle(\delta)$$

$$= \frac{4}{\epsilon^2} \int_s^\delta E\left[\frac{n}{R_n(u)} 1_{(R_n(u)>0)}\right] \mu(u) \, du$$

see (3.3). Now $R_n(u)$ is binomial with probability parameter $p = G(u)(1 - F(u))/\alpha$ and therefore the expectation in the integral above becomes

$$\sum_{k=1}^{n} n \binom{n}{k} \frac{1}{k} p^k (1 - p)^{n-k} \leq \frac{2}{p}$$

yielding via (3.5) the estimate

$$\Pr(\sup_{s,t:0<s\leq t\leq\delta} |X_n(s, t)| > \epsilon) \leq \frac{8}{\epsilon^2} \alpha \int_0^\delta \frac{\mu(u)}{G(u)(1 - F(u))} du$$

which is a bound that applies uniformly in n and by (3.4) tends to 0 as $\delta \downarrow 0$. We have established tightness close to 0.

Woodroofe's result (3.1) is local in the sense that it applied only to a restricted time domain. The global formulation (3.2) appears more satisfactory and in principle at least allows one to study the behavior of estimators of μ, even at timepoints close to 0.

REFERENCES

Billingsley, P. (1968). *Convergence of Probability Measures*. Wiley, New York.

Gill, R. D. (1984). A Note on two Papers in Central Limit Theory. Proc. 44th Session ISI, Madrid; *Bull. Inst. Internat. Statist.*, *50*, Vol. L, Book 3, 239–243.

Keiding, N., and Gill, R. D. (1990). *Random Truncation Models and Markov Processes*. Ann. Statist., *18*, 582–602.

Lindvall, T. (1973). Weak Convergence of Probability Measures and Random Functions in the Function Space $D[0, \infty)$. *J. Appl. Probab.*, *10*, 109–121.

Skorohod, A. V. (1956). Limit Theorems for Stochastic Processes. *Theory Probab. Appl.*, *1*, 261–290.

Whitt, W. (1980). Some Useful Functions for Functional Limit Theroems. *Math.Oper. Res.*, *5*, 67–85.

Woodroofe, M. (1985). Estimating a Distribution Function with Truncated Data. *Ann. Statist.*, *13*, 163–177.

7

Asymptotic Theory of Weighted Maximum Likelihood Estimation for Growth Models

B. L. S. Prakasa Rao
Indian Statistical Institute, New Delhi, India

Measurements such as height and weight are collected over a long period of time on an individual to study growth patterns. In general it might not always be possible to take the measurements at preassigned times and the times of observation may not be equally spaced. These observations, taken on the individual longitudinally, are dependent on each other. A new concept of weighted likelihood function corresponding to a weight function $\lambda(.)$ is introduced motivated by the fact that the times elapsed between observations should be taken into account in defining the likelihood function. The asymptotic properties of a weighted maximum likelihood estimator are investigated for parameters involved in studying such growth models. Some examples are given.

1 INTRODUCTION

Suppose measurements such as height and weight are collected longitudinally on an individual. In general, it is difficult to obtain the measurements at preassigned times of observation, for instance, due to the nonavailability of the subject itself. The times elapsed between different

183

times of observations might not be equal. The number of observations available on the individual at any time point is either none or at most one. Further the observations at different time points are possibly dependent on the earlier observations. For obvious reasons, the height at adolescent age of an individual depends on his height in childhood. Our interest is in the estimation of parameters involved in the stochastic modeling of such phenomena. An example of such a problem in the study of growth models is discussed in Section 4, Example 4.4 (see Pasternack and Shohoji (1976)).

As far as we are aware, the first work in the study of large sample properties of the maximum likelihood estimator in growth curves is due to Shohoji (1982) where he assumed that the observations are independent of each other. He mentioned that his result can be extended to the dependent case but no details were given. Here we consider a more general scheme of observations which are possibly dependent and study the asymptotic properties of a weighted maximum likelihood estimator corresponding to a weight function $\lambda(.)$. We assume that the observation scheme for the study is such that the maximum time elapsed between times of observation decreases to zero as the number of observations increases. The weighted function $\lambda(.)$ reflects the weightage given to the information obtained from the observations made at different times of observation.

For earlier work on the asymptotic properties of the maximum likelihood estimator for stochastic processes, see Basawa and Prakasa Rao (1980). In order to indicate the main ideas, we neither try to obtain the results under the weakest possible conditions nor do we discuss the multiparameter version. We will come back to these problems in a later publication.

Section 2 contains the definition of the *weighted maximum likelihood estimator* (WMLE) and the regularity conditions assumed. Asymptotic properties of WMLE are investigated in Section 3. It was shown that WMLE is consistent and asymptotic normal under the regularity conditions given in Section 2. A large number of examples are presented in Section 4. Some remarks about the choice of weight function are made in Section 5.

2 WEIGHTED MAXIMUM LIKELIHOOD ESTIMATION

Let Y_t be the observation recorded on an individual at time $t \in [a, b]$. Suppose the observations are obtained at time points

$$D_n: t_{0n} = a < t_{1n} < \cdots < t_{k_n n} = b \qquad (2.0)$$

For simplicity, we denote $Y_{t_{in}}$ by Y_{in}. We assume that $\{D_n, n \geq 1\}$ is a nested sequence. Let $\mathscr{F}_{ni}$ be the σ-algebra generated by $\{Y_{jn}, 1 \leq j \leq i\}$ for $1 \leq i \leq n$. Clearly $\mathscr{F}_{ni}$ is nondecreasing in i for each n. Since $\{D_n\}$ is nested, it follows that $\{\mathscr{F}_{ni}, 1 \leq i \leq n\}$ are nested in the sense that

$$\mathscr{F}_{ni} \subset \mathscr{F}_{n+1,i} \qquad \text{for } 1 \leq i \leq n$$

It is clear that the conditional distribution of the observation at time t_{in} depends on the particular time of observation and the earlier observations and their record times. Let $f(t_{in}, y_{in}, \beta \mid y_{jn}, t_{jn}: 1 \leq j \leq i - 1)$ be the conditional density of Y_{in} given the earlier observations assuming it exists. In order to take into account the time elapsed between measurements, we define the *weighted likelihood* corresponding to a weight function $\lambda(.)$ as

$$L_n(\beta) = \prod_{i=1}^{k_n} \{f(t_{in}, y_{in}, \beta \mid y_{jn}, t_{jn}: 1 \leq j \leq i - 1\}^{\lambda(t_{i,n}) - \lambda(t_{i-1,n})}$$

(2.1)

where $\lambda(.)$ is a known nondecreasing function on $[a, b]$. If $\lambda(t) \equiv t$ and $t_{in} - t_{i-1,n}$ is the same for all i, then we have a power of the usual likelihood function and the observations are made at equal time intervals. The motivation for defining the weighted likelihood function in this form is to give weightage to the information obtained at a particular time through the observation recorded at that time and at the time of observation.

The problem of interest is the estimation of β based on the observations $\{y_{1n}, \ldots, y_{k_n n}\}$ at times $\{t_{1n}, \ldots, t_{k_n n}\}$ respectively. $\hat{\beta}_{D_n}$ is called a WMLE if it maximizes $L_n(\beta)$ over the parameter space Ω of β. More precisely,

$$L_n(\hat{\beta}_{D_n}) = \sup_{\beta \in \Omega} L_n(\beta) \qquad (2.2)$$

Existence of a measurable weighted maximum likelihood estimator is assured if Ω is compact and the weighted likelihood function $L_n(\beta)$ is continuous in β and measurable in $(y_{1n}, \ldots, y_{k_n n}; t_{1n}, \ldots, t_{k_n n})$. We will not go into the details here. Hereafter we assume that there exists a (measurable) weighted maximum likelihood estimator of β.

Our aim is to study the asymptotic properties of the WMLE $\hat{\beta}_{D_n}$ as $\|D_n\| = \max_{1 \leq i \leq k_n} |t_{in} - t_{i-1,n}| \to 0$ as $n \to \infty$.

Without loss of generality, assume that the observations are made over the interval $[0, 1]$. Further assume that the following regularity conditions hold:

(A1) $\beta \in \Omega \subset R$ and Ω is an open interval.
(A2)   The conditional density functions

$$f(t_{in}, y_{in}; \beta \mid y_{jn}, t_{jn}; 1 \le j \le i-1) \equiv f_{in}$$

are twice differentiable with respect to β for almost all $(y_{1n}, \ldots, y_{k_n n})$, and all $\{t_{1n}, \ldots, t_{k_n n}\}$ in $[0, 1]$ and supports of these density functions do not depend on β.

(A3)   Differentiation twice with respect to β of

$$f_{in} \equiv f(t_{in}, y_{in}; \beta \mid y_{jn}, t_{jn}; 1 \le j \le i-1)$$

under the integral sign is allowed in

$$\int_{-\infty}^{\infty} f(t_{in}, y_{in}; \beta \mid y_{jn}, t_{jn}; 1 \le j \le i-1)\, d\mu = 1 \qquad (2.3)$$

where μ is a σ-finite measure on R with respect to which densities are obtained.

Assumption (A3) implies that

$$-E_\beta \left[\frac{d^2 \log f_{in}}{d\beta^2} \,\middle|\, y_{jn}, t_{jn}; 1 \le j \le i-1 \right]$$

$$= E_\beta \left[\left(\frac{d \log f_{in}}{d\beta} \right)^2 \,\middle|\, y_{jn}, t_{jn}; 1 \le j \le i-1 \right] \quad \text{a.s. } [\mu] \qquad (2.4)$$

The *weighted Fisher information* in the observations $(y_{in}, 1 \le i \le k_n)$ recorded at times $(t_{in}, 1 \le i \le k_n)$ is defined to be

$$I(D_n, \beta) \equiv -\sum_{i=1}^{k_n} [\lambda(t_{in}) - \lambda(t_{i-1,n})]$$

$$\times E_\beta \left[\frac{d^2 \log f_{in}}{d\beta^2} \,\middle|\, y_{jn}, t_{jn}, 1 \le j \le i-1 \right]. \qquad (2.5)$$

In view of (2.4), an alternate form for $I(D_n, \beta)$ is

$$I(D_n, \beta) = \sum_{i=1}^{k_n} [\lambda(t_{in}) - \lambda(t_{i-1,n})]$$

$$\times E_\beta \left[\left(\frac{d \log f_{in}}{d\beta} \right)^2 \,\middle|\, y_{jn}, t_{jn}, 1 \le j \le i-1 \right] \qquad (2.6)$$

which indicates that $I(D_n, \beta) \geq 0$ a.s. $[\mu]$. Note that $I(D_n, \beta)$ is possibly random. If $\{y_{jn}, 1 \leq j \leq k_n\}$ are independent and observations are made at equal intervals, $I(D_n, \beta)$ essentially reduces to the extended Fisher information discussed in Shohoji (1982).

In addition to the conditions (A1)–(A3), let us suppose that the following conditions hold:

(A4) There exists $J(\beta) > 0$ and nonrandom such that

(i) $\qquad |I(D_n, \beta) - J(\beta)| \xrightarrow{\; L_1\text{-mean} \;} 0 \quad$ and

(ii) $\qquad \left| \sum_{i=1}^{k_n} [\lambda(t_{in}) - \lambda(t_{i-1,n})] \dfrac{d^2 \log f_{in}}{d\beta^2} + J(\beta) \right| \xrightarrow{\; L_1\text{-mean} \;} 0$

as $\|D_n\| \to 0$.

(A5) There exists a sequence $\delta_n \to \infty$ and $\sigma^2(\beta) > 0$ nonrandom such that

(i) $\qquad \left| \delta_n^2 \sum_{i=1}^{k_n} [\lambda(t_{in}) - \lambda(t_{i-1,n})]^2 E_\beta \left(\dfrac{d \log f_{in}}{d\beta} \right)^2 \middle| y_{i-1,n} \right]$

$\qquad - \sigma^2(\beta) \Bigg| \xrightarrow{\; L_1\text{-mean} \;} 0$

and for every $\epsilon > 0$,

(ii) $\qquad \delta_n^2 \sum_{i=1}^{k_n} [\lambda(t_{in}) - \lambda(t_{i-1,n})]^2 E_\beta \left[\left\{ \left(\dfrac{d \log f_{in}}{d\beta} \right)^2 \right. \right.$

$\qquad \left. \left. \times \chi \left(\left| \delta_n \dfrac{d \log f_{in}}{d\beta} \{\lambda(t_{in}) - \lambda(t_{i-1,n})\} \right| > \epsilon \right) \middle| y_{i-1,n} \right] \xrightarrow{\; P \;} 0$

as $\|D_n\| \to 0$ where $\chi(A)$ denotes the indicator function of set A and $y_{i-1,n}$ denotes the set $(y_{j,n}, t_{jn}: 1 \leq j \leq i-1)$.

Note that condition (A4) implies that

$$ E_\beta \left(-\sum_{i=1}^{k_n} [\lambda(t_{in}) - \lambda(t_{i-1,n})] \dfrac{d^2 \log f_{in}}{d\beta^2} \right) = E_\beta(I(D_n, \beta)) \to J(\beta) \quad (2.7) $$

as $\|D_n\| \to 0$.

3 ASYMPTOTIC PROPERTIES

We first discuss conditions under which a WMLE is consistent. Suppose that the regularity conditions (A1)–(A5) hold.

Let β_0 denotes the true parameter. Expanding $d \log L_n/d\beta$ around β_0, we have

$$
\begin{aligned}
\frac{d \log L_n}{d\beta} &= \frac{d \log L_n}{d\beta}\bigg|_{\beta_0} + (\beta - \beta_0) \frac{d^2 \log L_n}{d\beta^2}\bigg|_{\beta=\beta^*} \\
&= \frac{d \log L_n}{d\beta}\bigg|_{\beta_0} - (\beta - \beta_0)I(D_n, \beta_0) + (\beta - \beta_0) \\
&\quad \times \left[\frac{d^2 \log L_n}{d\beta^2}\bigg|_{\beta=\beta^*} + I(D_n, \beta_0) \right]
\end{aligned}
\tag{3.0}
$$

Observe that the probability that $I(D_n, \beta_0) > 0$ approaches 1 as $\|D_n\| \to 0$ when β_0 is the true parameter in view of (A4).

Consider the sequence

$$
\frac{\dfrac{d \log L_n}{d\beta}\bigg|_{\beta_0}}{I(D_n, \beta_0)} = \frac{\displaystyle\sum_{i=1}^{k_n} [\lambda(t_{in}) - \lambda_{(i-1,n)}] \dfrac{d \log f_{in}}{d\beta}\bigg|_{\beta_0}}{I(D_n, \beta_0)}
\tag{3.1}
$$

Note that

$$
E_{\beta_0}\left(\frac{d \log L_n(\beta)}{d\beta}\bigg|_{\beta_0} \right) = 0
$$

from the definition of $L_n(\beta)$ and (A3). Hence

$$
\begin{aligned}
E_{\beta_0}\left(\frac{d \log L_n(\beta)}{d\beta}\bigg|_{\beta_0} \right)^2 &= \mathrm{Var}_{\beta_0}\left(\frac{d \log L_n(\beta)}{d\beta}\bigg|_{\beta_0} \right) \\
&= \sum_{i=1}^{k_n} [\lambda(t_{in}) - \lambda(t_{i-1,n})]^2 E_{\beta_0}\left(\frac{d \log f_{in}}{d\beta}\bigg|_{\beta_0} \right)^2
\end{aligned}
$$

and the last term tends to zero as $\|D_n\| \to 0$ by (A5)(i). Therefore the numerator in (3.1) tends to zero in probability. The denominator in (3.1) tends to $J(\beta_0) > 0$ in probability by assumption (A4). Hence the ratio in (3.1) tends to zero in probability and

$$\frac{d \log L_n}{d\beta} [I(D_n, \beta_0)]^{-1} =$$

$$\frac{d \log L_n}{d\beta}\bigg|_{\beta_0} [I(D_n, \beta_0)]^{-1} - (\beta - \beta_0)$$

$$+ (\beta - \beta_0) \left[\frac{d^2 \log L_n}{d\beta^2}\bigg|_{\beta^*} + I(D_n, \beta_0) \right] [I(D_n, \beta_0)]^{-1}$$

$$= o_p(1) - (\beta - \beta_0) + (\beta - \beta_0) \left[\frac{d^2 \log L_n}{d\beta^2}\bigg|_{\beta^*} + I(D_n, \beta_0) \right]$$

$$\times [I(D_n, \beta_0)]^{-1} \tag{3.2}$$

Suppose the following condition holds in addition to regularity conditions (A1)–(A5) stated in Section 2.

(A6) $$\varprojlim_{\|D_n\| \to 0} \left| \frac{\left| \frac{d^2 \log L_n}{d\beta^2}\bigg|_{\beta=\beta^*} + I(D_n, \beta_0) \right|}{I(D_n, \beta_0)} \right| < 1 \text{ in probability}$$

Then it follows that the equation

$$\frac{d \log L_n}{d\beta} = 0$$

has a solution which is weakly consistent. One can give alternative versions of sufficient conditions which ensure weak consistency following Basawa and Prakasa Rao (1980). However we will not go into the discussion here. Our approach here is similar to Hall and Heyde (1980), p. 159. We have the following result from the above remarks.

Theorem 3.1. Under the conditions (A1) to (A6), there exists a weakly consistent solution of the likelihood equation with probability approaching one as $\|D_n\| \to 0$.

Let us now study asymptotic normality of a weighted maximum likelihood estimator.

In addition to the conditions (A1) to (A5), let us assume that

(A7) $\hat{\beta}_{D_n}$ is a weakly consistent estimator of β_0; and

(A8)   suppose that the conditional density function

$$f_{in} = f(t_{in}, y_{in}; \beta \mid y_{jn}; 1 \le j \le i-1)$$

is thrice differentiable with respect to β for almost all $(y_{1n}, \ldots, y_{in})$ and there exists a neighborhood V_{β_0} for every $\beta_0 \in \Omega$ such that

$$\sup_{\beta \in V_{\beta_0}} \left| \frac{d^3 \log f_{in}}{d\beta^3} \right| = O_p(1) \tag{3.3}$$

Note that

$$L_n(\beta) = \prod_{i=1}^{k_n} \{f(t_{in}, y_{in}; \beta \mid y_{jn}, t_{jn}; 1 \le j \le i-1)\}^{\lambda(t_{in}) - \lambda(t_{i-1,n})} \tag{3.4}$$

and hence

$$\log L_n(\beta) = \sum_{i=1}^{k_n} [\lambda(t_{i,n}) - \lambda(t_{i-1,n})] \log f_{in} \tag{3.5}$$

Since $\hat{\beta}_{D_n}$ is weakly consistent, the probability that $\hat{\beta}_{D_n}$ lies in a neighborhood of β_0 tends to one as $\|D_n\| \to 0$. Expanding $d \log L_n(\beta)/d\beta$ around β_0, we have

$$\frac{d \log L_n(\beta)}{d\beta} = \sum_{i=1}^{k_n} [\lambda(t_{in}) - \lambda(t_{i-1,n})] \frac{d \log f_{in}}{d\beta} \Big|_{\beta=\beta_0}$$

$$+ (\beta - \beta_0) \sum_{i=1}^{k_n} [\lambda(t_{in}) - \lambda(t_{i-1,n})] \frac{d^2 \log f_{in}}{d\beta^2} \Big|_{\beta=\beta_0}$$

$$+ \frac{(\beta - \beta_0)^2}{2} \sum_{i=1}^{k_n} [\lambda(t_{in}) - \lambda(t_{i-1,n})] \frac{d^3 \log f_{in}}{d\beta^3} \Big|_{\beta=\beta^*} \tag{3.6}$$

where $|\beta^* - \beta_0| \le |\beta - \beta_0|$. In particular, for $\beta = \hat{\beta}_{D_n}$, it follows that

$$0 = \frac{d \log L_n(\beta)}{d\beta} \Big|_{\beta=\hat{\beta}_{D_n}} = \sum_{i=1}^{k_n} [\lambda(t_{in}) - \lambda(t_{i-1,n})] \frac{d \log f_{in}}{d\beta} \Big|_{\beta=\beta_0}$$

$$+ (\hat{\beta}_{D_n} - \beta_0) \sum_{i=1}^{k_n} [\lambda(t_{in}) - \lambda(t_{i-1,n})] \frac{d^2 \log f_{in}}{d\beta^2} \Big|_{\beta=\beta_0}$$

$$+ \frac{(\hat{\beta}_{D_n} - \beta_0)^2}{2} \sum_{i=1}^{k_n} [\lambda(t_{in}) - \lambda(t_{i-1,n})] \frac{d^3 \log f_{in}}{d\beta^3} \Big|_{\beta=\beta^*_{D_n}} \tag{3.7}$$

where $|\beta_{D_n}^* - \beta_0| \le |\hat{\beta}_{D_n} - \beta_0|$. Therefore

$$\hat{\beta}_{D_n} - \beta_0 =$$

$$= \frac{\sum_{i=1}^{k_n} [\lambda(t_{in}) - \lambda(t_{i-1,n})] \frac{d \log f_{in}}{d\beta}\bigg|_{\beta=\beta_0}}{-\sum_{i=1}^{k_n} [\lambda(t_{in}) - \lambda(t_{i-1,n})] \frac{d^2 \log f_{in}}{d\beta^2}\bigg|_{\beta=\beta_0} - \frac{(\hat{\beta}_{D_n} - \beta_0)}{2} \sum_{i=1}^{k_n} [\lambda(t_{in}) - \lambda(t_{i-1,n})] \frac{d^3 \log f_{in}}{d\beta^3}\bigg|_{\beta=\beta_{D_n}^*}}$$

$$(3.8)$$

Hence

$$\delta_n(\hat{\beta}_{D_n} - \beta_0) = \frac{\sum_{i=1}^{k_n} \delta_n[\lambda(t_{in}) - \lambda(t_{i-1,n})] \frac{d \log f_{in}}{d\beta}\bigg|_{\beta_0}}{\sum_{i=1}^{k_n} [\lambda(t_{in}) - \lambda(t_{i-1,n})] \frac{d^2 \log f_{in}}{d\beta^2}\bigg|_{\beta=\beta_0}} + o_p(1)$$

$$(3.9)$$

by assumptions (A7) and (A8). Assumption (A5) implies

$$\sum_{i=1}^{k_n} \delta_n[\lambda(t_{in}) - \lambda(t_{i-1,n})] \frac{d \log f_{in}}{d\beta}\bigg|_{\beta_0} \xrightarrow{\mathcal{L}} N(0, \sigma^2(\beta_0)) \qquad (3.10)$$

by Corollary 3.1, p. 58 in Hall and Heyde (1980). Condition (A4) implies that

$$-\sum_{i=1}^{k_n} [\lambda(t_{in}) - \lambda(t_{i-1,n})] \frac{d^2 \log f_{in}}{d\beta^2}\bigg|_{\beta=\beta_0} \xrightarrow{p} J(\beta_0) \qquad (3.11)$$

as $\|D_n\| \to 0$. Furthermore the second term in the denominator on the right-hand side of (3.9) is $o_p(1)$ from assumption (A4). Combining all these obervations, it is easy to check that

$$\delta_n(\hat{\beta}_{D_n} - \beta_0) \xrightarrow{\mathcal{L}} N\left(0, \frac{\sigma^2(\beta_0)}{J^2(\beta_0)}\right) \quad \text{as} \quad \|D_n\| \to 0$$

Theorem 3.2.   Under the condition (A1) to (A5), (A7) and (A8),

$$\delta_n(\hat{\beta}_{D_n} - \beta_0) \xrightarrow{\mathcal{L}} N\left(0, \frac{\sigma^2(\beta_0)}{J^2(\beta_0)}\right) \quad \text{as} \quad \|D_n\| \to 0$$

Remark 3.1.   If $\lambda(t) = t$, $0 \le t \le 1$, $k_n = n$ and $t_{in} - t_{i-1,n} = 1/n$ for all

$1 \le i \le n$, then δ_n can usually be chosen as $n^{1/2}$ in case $Y_{in}, i = 1, \ldots, n$ are independent. The condition (A4) reduces to

$$\left| \frac{1}{n} \sum_{i=1}^{n} E_\beta \left(\frac{d^2 \log f_{in}}{d\beta^2} \right) + \sigma^2(\beta) \right| \to 0$$

and (A5) reduces to

$$\left| \frac{1}{n} \sum_{i=1}^{n} E_\beta \left(\frac{d \log f_{in}}{d\beta} \right)^2 - J(\beta) \right| \to 0$$

But

$$-E_\beta \left[\frac{d^2 \log f_{in}}{d\beta^2} \right] = E_\beta \left[\frac{d \log f_{in}}{d\beta} \right]^2$$

and hence $\sigma^2(\beta_0) = J(\beta_0)$ and we have

$$n^{1/2}(\hat{\beta}_{Dn} - \beta_0) \xrightarrow{\mathcal{L}} N \left(0, \frac{1}{J(\beta_0)} \right) \quad \text{as} \quad \|D_n\| \to 0$$

Remark 3.2. The problem of nonergodic model (see Basawa and Scott (1983)) is interesting here in the sense that the function $J(\beta)$ and $\sigma^2(\beta)$ could be possibly random. It should be possible to handle this case using techniques in Proposition 6.1 of Hall and Heyde (1980), p. 160. We will come back to this problem in another publication.

4 EXAMPLES

We have developed general results on asymptotic properties of weighted MLEs under some regularity conditions. Most often it is easy to derive the properties of WMLE from the special structure in the model rather than verify the regularity conditions. We shall now discuss some examples.

Example 4.1. Suppose $Y_t \sim N(\beta t, 1)$, $0 \le t \le 1$. The problem is to estimate β based on the observation at times $0 = t_{0n} < t_{1n} < \cdots < t_{k_n n} = 1$. We assume that there is only one observation at each time point and Y_t are independent. The weighted likelihood is

$$L_n(\beta) = \left(\frac{1}{\sqrt{2\pi}} \right)^{k_n} \prod_{i=1}^{k_n} \{ e^{-1/2(y_{t_{i-1,n}} - \beta t_{i-1,n})^2} \}^{\lambda(t_{in}) - \lambda(t_{i-1,n})} \qquad (4.0)$$

and hence

$$\frac{d \log L_n(\beta)}{d\beta} = \sum_{i=1}^{k_n} [\lambda(t_{in}) - \lambda(t_{i-1,n})](y_{t_{i-1,n}} - \beta t_{i-1,n})t_{i-1,n}$$

Hence the WMLE $\hat{\beta}_{D_n}$ is given by

$$\hat{\beta}_{D_n} = \frac{\sum_{i=1}^{k_n} y_{t_{i-1,n}} t_{i-1,n}[\lambda(t_{in}) - \lambda(t_{i-1,n})]}{\sum_{i=1}^{k_n} t_{i-1,n}^2 [\lambda(t_{in}) - \lambda(t_{i-1,n})]} \tag{4.1}$$

It is easy to see that $\hat{\beta}_{D_n}$ maximizes $L_n(\beta)$. Observe that $\hat{\beta}_{D_n}$ is an unbiased estimator of β.

In fact

$$\hat{\beta}_{D_n} - \beta = \frac{\sum_{i=1}^{k_n} (y_{t_{i-1,n}} - \beta t_{i-1,n})t_{i-1,n}[\lambda(t_{in}) - \lambda(t_{i-1,n})]}{\sum_{i=1}^{k_n} t_{i-1,n}^2 [\lambda(t_{in}) - \lambda(t_{i-1,n})]} \tag{4.2}$$

Therefore

$$\begin{aligned}
E|\hat{\beta}_{D_n} - \beta|^2 = \text{Var}(\hat{\beta}_{D_n}) &= \frac{\sum_{i=1}^{k_n} \text{Var}(y_{t_{i-1,n}})t_{i-1,n}^2 [\lambda(t_{in}) - \lambda(t_{i-1,n})]^2}{\left(\sum_{i=1}^{k_n} t_{i-1,n}^2 \{\lambda(t_{in}) - \lambda(t_{i-1,n})\}\right)^2} \\
&= \frac{\sum_{i=1}^{k_n} t_{i-1,n}^2 [\lambda(t_{in}) - \lambda(t_{i-1,n})]^2}{\left(\sum_{i=1}^{k_n} t_{i-1,n}^2 \{\lambda(t_{in}) - \lambda(t_{i-1,n})\}\right)^2} \tag{4.3}
\end{aligned}$$

Note that if $\lambda(.)$ is such that

$$\frac{\sum_{i=1}^{k_n} t_{i-1,n}^2 [\lambda(t_{in}) - \lambda(t_{i-1,n})]^2}{\left(\sum_{i=1}^{k_n} t_{i-1,n}^2 \{\lambda(t_{in}) - \lambda(t_{i-1,n})\}\right)^2} \to 0 \quad \text{as} \quad \|D_n\| \to 0$$

then $E|\hat{\beta}_{D_n} - \beta|^2 \to 0$ as $n \to \infty$ and hence $\hat{\beta}_{D_n}$ is a mean square consistent estimator. For instance if $\lambda(t) = t$, then $k_n = n$ and

$$\sum_{i=1}^{n} t_{i-1,n}^2(t_{in} - t_{i-1,n}) \to \int_0^1 t^2 \, dt = \frac{1}{3} \quad \text{as} \quad \|D_n\| \to 0$$

and

$$\sum_{i=1}^{n} t_{i-1,n}^2(t_{in} - t_{i-1,n})^2 \leq \max_{1 \leq i \leq n} |t_{in} - t_{i-1,n}|$$

$$\times \sum_{i=1}^{n} t_{i-1,n}^2(t_{in} - t_{i-1,n}) \to 0$$

Hence $\hat{\beta}_{Dn} \to \beta$ in L_2-mean as $n \to \infty$.

It follows from (4.2) that

$$(\hat{\beta}_{Dn} - \beta)\left(\sum_{i=1}^{k_n} t_{i-1,n}^2\{\lambda(t_{in}) - \lambda(t_{i-1,n})\}\right)$$

$$= \sum_{i=1}^{k_n} (y_{t_{i-1,n}} - \beta t_{i-1,n})t_{i-1,n}\{\lambda(t_{in}) - \lambda(t_{i-1,n})\} \tag{4.4}$$

Note that the expression on the right side has exact

$$N\left(0, \sum_{i=1}^{k_n} t_{i-1,n}^2[\lambda(t_{in}) - \lambda(t_{i-1,n})]^2\right)$$

Hence

$$(\hat{\beta}_{Dn} - \beta) \text{ has } N\left(0, \frac{\displaystyle\sum_{i=1}^{k_n} t_{i-1,n}^2[\lambda(t_{in}) - \lambda(t_{i-1,n})]^2}{\left(\displaystyle\sum_{i=1}^{k_n} t_{i-1,n}^2\{\lambda(t_{in}) - \lambda(t_{i-1,n})\}\right)^2}\right) \tag{4.5}$$

Example 4.2. Suppose $Y_t \sim N(\theta + t^{-\alpha}, 1)$, $0 \leq t \leq 1$ under the same setup as in Example 4.1 where α is known. The problem is to estimate θ. Let

$$D_n: 0 = t_{on} < t_{1n} < \cdots < t_{k_n n} = 1 \tag{4.6}$$

be a subdivision of $[0, 1]$. Suppose Y_t are independent as before. Then

$$L_n(\theta) = \sum_{i=1}^{k_n} \left\{\frac{1}{\sqrt{2\pi}} \exp\left[-\frac{1}{2}(y_{t_{i-1,n}} - \theta - t_{i-1,n}^{-\alpha})^2\right]\right\}^{\lambda(t_{in}) - \lambda(t_{i-1,n})} \tag{4.7}$$

and hence

$$\frac{d \log L_n(\theta)}{d\theta} = (-\tfrac{1}{2}) \sum_{i=1}^{k_n} 2(y_{t_{i-1,n}} - \theta - t_{i-1,n}^{-\alpha})\{\lambda(t_{in}) - \lambda(t_{i-1,n})\}$$

Therefore

$$\hat{\theta}_n = \frac{\displaystyle\sum_{i=1}^{k_n} (y_{t_{i-1,n}} - t_{i-1,n}^{-\alpha})\{\lambda(t_{in}) - \lambda(t_{i-1,n})\}}{\displaystyle\sum_{i=1}^{k_n} \{\lambda(t_{in}) - \lambda(t_{i-1,n})\}}$$

$$= \frac{\displaystyle\sum_{i=1}^{k_n} (y_{t_{i-1,n}} - t_{i-1,n}^{-\alpha})(\lambda(t_{in}) - \lambda(t_{i-1,n}))}{\lambda(1) - \lambda(0)} \qquad (4.8)$$

It is easy to check that $\hat{\theta}_n$ is a WMLE since

$$\left.\frac{d^2 \log L_n(\theta)}{d\theta^2}\right|_{\theta=\hat{\theta}} < 0$$

Now

$$\hat{\theta}_n - \theta = \sum_{i=1}^{k_n} (y_{t_{i-1,n}} - \theta - t_{i-1,n}^{-\alpha})$$

$$\times \{\lambda(t_{in}) - \lambda(t_{i-1,n})\} \Big/ \sum_{i=1}^{k_n} [\lambda(t_{in}) - \lambda(t_{i-1,n})]$$

and

$$E|\hat{\theta}_n - \theta|^2 = \sum_{i=1}^{k_n} [\lambda(t_{in}) - \lambda(t_{i-1,n})]^2/[\lambda(1) - \lambda(0)]^2$$

Hence $\hat{\theta}_n$ is mean square consistent if

$$\sum_{i=1}^{k_n} [\lambda(t_{in}) - \lambda(t_{i-1,n})]^2 \to 0 \quad \text{as } \|D_n\| \to 0$$

Furthermore

$$\hat{\theta}_n - \theta \text{ has } N\left(0, \sum_{i=1}^{k_n} [\lambda(t_{in}) - \lambda(t_{i-1,n})]^2\right) \qquad (4.9)$$

Example 4.3. Suppose $Y_t = \eta t + B(t)$, $0 \le t \le 1$ where $\{B(T),$

$0 \leq t \leq 1\}$ is standard Brownian motion. Suppose the process is observed at time points

$$t_{0,n} = 0 < t_{1,n} < \cdots < t_{k_n,n} = 1 \qquad (4.10)$$

It is clear that the conditional distribution of $Y_{t_{i,n}}$ given $Y_{t_{1,n}}, \ldots, Y_{t_{i-1,n}}$ depends only on $Y_{t_{i-1,n}}$ by the fact that the increments of the Wiener process are independent. Hence the weighted likelihood is

$$L_n(\eta) = \prod_{i=1}^{k_n} \left\{ \frac{1}{\sqrt{2\pi(t_{i,n} - t_{i-1,n})}} \right.$$

$$\left. \times e^{-1/2 \frac{[Y_{t_{i,n}} - Y_{t_{i-1,n}} - \eta(t_{i,n} - t_{i-1,n})]^2}{t_{i,n} - t_{i-1,n}}} \right\}^{\lambda(t_{i,n}) - \lambda(t_{i-1,n})}$$

$$(4.11)$$

Therefore

$$\frac{d \log L_n(\eta)}{d\eta} = \sum_{i=1}^{k_n} [Y_{t_{i,n}} - Y_{t_{i-1,n}} - \eta(t_{in} - t_{i-1,n})]\{\lambda(t_{i,n}) - \lambda(t_{i-1,n})\}$$

and

$$\hat{\eta}_{D_n} = \frac{\displaystyle\sum_{i=1}^{k_n} (Y_{t_{i,n}} - Y_{t_{i-1,n}})\{\lambda(t_{i,n}) - \lambda(t_{i-1,n})\}}{\displaystyle\sum_{i=1}^{k_n} (t_{in} - t_{i-1,n})\{\lambda(t_{in}) - \lambda(t_{i-1,n})\}} \qquad (4.12)$$

is a WMLE of η. It is easy to see again that

$$\left. \frac{d^2 \log L_n(\eta)}{d\eta^2} \right|_{\eta = \hat{\eta}_{D_n}} < 0$$

Furthermore

$$\hat{\eta}_{D_n} - \eta = \frac{\displaystyle\sum_{i=1}^{k_n} [(Y_{t_{i,n}} - Y_{t_{i-1,n}}) - \eta(t_{in} - t_{i-1,n})]\{\lambda(t_{in}) - \lambda(t_{i-1,n})\}}{\displaystyle\sum_{i=1}^{k_n} (t_{in} - t_{i-1,n})\{\lambda(t_{in}) - \lambda(t_{i-1,n})\}}$$

$$= \frac{\displaystyle\sum_{i=1}^{k_n} [B(t_{i,n}) - B(t_{i-1,n})]\{\lambda(t_{in}) - \lambda(t_{i-1,n})\}}{\displaystyle\sum_{i=1}^{k_n} (t_{in} - t_{i-1,n})\{\lambda(t_{in}) - \lambda(t_{i-1,n})\}}$$

Note that $\hat{\eta}_{Dn}$ is unbiased estimator of η. Furthermore

$$E|\hat{\eta}_{Dn} - \eta|^2 = \frac{\displaystyle\sum_{i=1}^{k_n} [\lambda(t_{in}) - \lambda(t_{i-1,n})]^2 (t_{in} - t_{i-1,n})}{\left[\displaystyle\sum_{i=1}^{k_n} (t_{in} - t_{i-1,n})\{\lambda(t_{in}) - \lambda(t_{i-1,n})\}\right]^2}$$

In fact

$$\hat{\eta}_{D_n} - \eta \text{ is } N\left(0, \frac{\displaystyle\sum_{i=1}^{k_n} (t_{in} - t_{i-1,n})[\lambda(t_{in}) - \lambda(t_{i-1,n})]^2}{\left[\displaystyle\sum_{i=1}^{k_n} (t_{in} - t_{i-1,n})\{\lambda(t_{in}) - \lambda(t_{i-1,n})\}\right]^2}\right) \quad (4.13)$$

If the data is equally spaced and $\lambda(t) \equiv t$ than $\hat{\eta}_{D_n} - \eta$ is $N(0, 1)$ and $\hat{\eta}_{D_n}$ is not consistent. In general

$$\frac{\displaystyle\sum_{i=1}^{k_n} (t_{in} - t_{i-1,n})[\lambda(t_{in}) - \lambda(t_{i-1,n})]^2}{\left[\displaystyle\sum_{i=1}^{k_n} (t_{in} - t_{i-1,n})\{\lambda(t_{in}) - \lambda(t_{i-1,n})\}\right]^2} \geq 1$$

by the Cauchy-Schwartz inequality, equality occurring iff $\lambda(t_{in}) - \lambda(t_{i-1,n})$ is the same for all $1 \leq i \leq n$. Hence the spread of $\hat{\eta}_{D_n}$ is wider in this general case and the estimator $\hat{\eta}_{D_n}$ is not consistent no matter what $\lambda(.)$ is and however fine the subdivision D_n of $[0, 1]$ is.

Example 4.4. Suppose Y_t is $N(h(t, \gamma), 1)$, $0 \leq t \leq 1$ where $h(t, \gamma)$ is a known function but for γ. The problem is to estimate γ based on the observations made at times

$$0 = t_{0,n} < t_{1,n} < \cdots < t_{k_n,n} = 1 \quad (4.14)$$

We assume as before that there is one and only one observation recorded at each time point and Y_t are independent. The likelihood function here is proportional to

$$L_n(\gamma) = \prod_{i=1}^{k_n} \{e^{-(1/2)(Y_{t_{i-1,n}} - h(t_{i-1,n}, \gamma))^2}\}^{[\lambda(t_{i,n}) - \lambda(t_{i-1,n})]} \quad (4.15)$$

and hence

$$\log L_n(\gamma) = -\frac{1}{2} \sum_{i=1}^{k_n} (Y_{t_{i-1,n}} - h(t_{i-1,n}, \gamma))^2 \{\lambda(t_{i,n}) - \lambda(t_{i-1,n})\}$$

This implies that

$$\frac{d \log L_n(\gamma)}{d\gamma} = \sum_{i=1}^{k_n} (Y_{t_{i-1,n}} - h(t_{i-1,n}, \gamma)) \frac{\partial}{\partial \gamma} h(t_{i-1,n}, \gamma)\{\lambda(t_{i,n}) - \lambda(t_{i-1,n})\}$$

Hence the likelihood equation can be written in the form

$$\sum_{i=1}^{k_n} Y_{t_{i-1,n}} \frac{\partial}{\partial \gamma} h(t_{i-1,n}, \hat{\gamma})(\lambda(t_{i,n}) - \lambda(t_{i-1,n}))$$

$$= \sum_{i=1}^{k_n} h(t_{i-1,n}, \hat{\gamma}) \frac{\partial}{\partial \gamma} h(t_{i-1,n}, \hat{\gamma})\{\lambda(t_{i,n}) - \lambda(t_{i-1,n})\} \qquad (4.16)$$

where $\hat{\gamma}$ denotes a solution of the likelihood equation. In particular if an MLE of γ exists, then it is a solution of the above equation. Let us now consider

$$\frac{d^2 \log L_n(\gamma)}{d\gamma^2} = \sum_{i=1}^{k_n} (Y_{t_{i-1,n}} - h(t_{i-1,n}, \gamma)) \frac{\partial^2}{\partial \gamma^2} h(t_{i-1,n}, \gamma)\{\lambda(t_{i,n}) - \lambda(t_{i-1,n})\}$$

$$- \sum_{i=1}^{k_n} \left[\frac{\partial h(t_{i-1,n}, \gamma)}{\partial \gamma}\right]^2 \{\lambda(t_{i,n}) - \lambda(t_{i-1,n})\}$$

Note that

$$E_\gamma\left[\frac{d^2 \log L_n(\gamma)}{d\gamma^2}\right] = - \sum_{i=1}^{k_n} \left[\frac{\partial h(t_{i-1,n}, \gamma)}{\partial \gamma}\right]^2 [\lambda(t_{i,n}) - \lambda(t_{i-1,n})]$$

and the last term tends to

$$- \int_0^1 \left[\frac{\partial h(t, \gamma)}{\partial \gamma}\right]^2 d\lambda(t)$$

It is easy to see from the preceding computation that the weighted Fisher information is

$$I(D_n, \gamma) = \sum_{i=1}^{k_n} [\lambda(t_{i,n}) - \lambda(t_{i-1,n})]^2 \left[\frac{\partial}{\partial \gamma} h(t_{i-1,n}, \gamma)\right]^2$$

$$\rightarrow \int_0^1 \left[\frac{\partial}{\partial \gamma} h(t, \gamma)\right]^2 d\lambda(t) \quad \text{as } \|D_n\| \rightarrow 0 \qquad (4.17)$$

Hence

$$J(\gamma) = \int_0^1 \left[\frac{\partial}{\partial \gamma} h(t, \gamma)\right]^2 d\lambda(t) \qquad (4.18)$$

On the other hand,

$$\sum_{i=1}^{k_n} [\lambda(t_{in}) - \lambda(t_{i-1,n})]^2 E_\gamma \left[\left(\frac{d \log f_{in}}{d\gamma} \right)^2 \right]$$

$$= \sum_{i=1}^{k_n} [\lambda(t_{in}) - \lambda(t_{i-1,n})]^2 \operatorname{Var}_\gamma \left[\frac{d \log f_{in}}{d\gamma} \right]$$

$$= \sum_{i=1}^{k_n} [\lambda(t_{in}) - \lambda(t_{i-1,n})]^2 \left[\frac{\partial}{\partial\gamma} h(t_{i-1,n}, \gamma) \right]^2 \qquad (4.19)$$

Suppose there exists $\delta_n \to \infty$ such that

$$\delta_n^2 \sum_{i=1}^{k_n} [\lambda(t_{in}) - \lambda(t_{i-1,n})]^2 \left[\frac{\partial}{\partial\gamma} h(t_{i-1,n}, \gamma) \right]^2 \to \sigma^2(\gamma)$$

as $\|D_n\| \to 0$. $\qquad (4.20)$

Under the above condition, it follows that

$$\delta_n(\hat{\gamma}_{D_n} - \gamma) \overset{\mathcal{L}}{\longrightarrow} N\left(0, \frac{\sigma^2(\gamma)}{J^2(\gamma)} \right) \quad \text{as } \|D_n\| \to 0 \qquad (4.21)$$

where $\hat{\gamma}_{D_n}$ is WMLE of γ. Note that $\sigma^2(\gamma)$ and $J^2(\gamma)$ depend on the weight function λ.

Suppose $t_{in} - t_{i-1,n} = \Delta_n$ for $1 \le i \le k_n$. Then, assuming that λ is continuously differentiable, we have

$$\sum_{i=1}^{k_n} [\lambda(t_{in}) - \lambda(t_{i-1,n})]^2 \left[\frac{\partial}{\partial\gamma} h(t_{i-1,n}, \gamma) \right]^2$$

$$\approx \Delta_n \sum_{i=1}^{k_n} [\lambda'(t_{i-1,n})]^2 \left[\frac{\partial}{\partial\gamma} h(t_{i-1,n}, \gamma) \right]^2 (t_{in} - t_{i-1,n})$$

$$\approx \Delta_n \int_0^1 [\lambda'(t)]^2 \left[\frac{\partial}{\partial\gamma} h(t, \gamma) \right]^2 dt \qquad (4.22)$$

and hence Δ_n can be chosen to be δ_n^{-2} and

$$\delta_n^2 \sum_{i=1}^{k_n} [\lambda(t_{in}) - \lambda(t_{i-1,n})]^2 \left[\frac{\partial}{\partial\gamma} h(t_{i-1,n}, \gamma) \right]^2$$

$$= \Delta_n^{-1} \sum_{i=1}^{k_n} [\lambda(t_{in}) - \lambda(t_{i-1,n})]^2 \left[\frac{\partial}{\partial\gamma} h(t_{i-1,n}, \gamma) \right]^2$$

$$\to \int_0^1 [\lambda'(t)]^2 \left[\frac{\partial}{\partial\gamma} h(t, \gamma) \right]^2 dt \equiv \sigma^2(\gamma) \qquad (4.23)$$

as $\|D_n\| \to 0$. In this case,

$$J(\gamma) = \int_0^1 \left[\frac{\partial}{\partial \gamma} h(t, \gamma) \right]^2 \lambda'(t)\, dt \qquad (4.24)$$

Therefore

$$\Delta_n^{-1/2}(\hat{\gamma}_{D_n} - \gamma) \xrightarrow{\mathscr{L}} N\left(0, \frac{\int_0^1 [\lambda'(t)]^2 \left[\frac{\partial}{\partial \gamma} h(t, \gamma) \right]^2 dt}{\left\{ \int_0^1 \lambda'(t) \left[\frac{\partial}{\partial \gamma} h(t, \gamma) \right]^2 dt \right\}^2}\right) \qquad (4.25)$$

If $\lambda(t) \equiv t$, then

$$\Delta_n^{-1/2}(\hat{\gamma}_{D_n} - \gamma) \xrightarrow{\mathscr{L}} N\left[0, \frac{1}{\int_0^1 \left[\frac{\partial}{\partial \gamma} h(t, \gamma) \right]^2 dt}\right] \qquad (4.26)$$

The optimal weight function $\lambda(.)$ is obtained by minimizing the asymptotic variance

$$\frac{\int_0^1 [\lambda'(t)]^2 \left[\frac{\partial}{\partial \gamma} h(t, \gamma) \right]^2 dt}{\left\{ \int_0^1 \lambda'(t) \left[\frac{\partial}{\partial \gamma} h(t, \gamma) \right]^2 dt \right\}^2} \qquad (4.27)$$

over all $\lambda(.)$ nonnegative increasing differentiable functions on $[0, 1]$ with fixed $\lambda(0) = 0$ and $\lambda(1) = 1$ (say). By the Cauchy–Schwartz inequality

$$\left(\int_0^1 \lambda'(t) \left[\frac{\partial}{\partial \gamma} h(t, \gamma) \right]^2 dt \right)^2$$

$$\leq \int_0^1 [\lambda'(t)]^2 \left[\frac{\partial}{\partial \gamma} h(t, \gamma) \right]^2 dt \int_0^1 \left[\frac{\partial}{\partial \gamma} h(t, \gamma) \right]^2 dt \qquad (4.28)$$

equality occurring iff

$$\lambda'(t) \frac{\partial}{\partial \gamma} [h(t, \gamma)] \quad \text{and} \quad \frac{\partial}{\partial \gamma} [h(t, \gamma)]$$

are linearly related. In other words $\lambda'(t)$ has to be constant a.s. on

[0, 1]. This in turn proves that $\lambda(t) = \alpha t + \beta$ for some α and β. Since $\lambda(t)$ has to be nonnegative with $\lambda(0) = 0$ and $\lambda(1) = 1$, we get that $\lambda(t) = t$. For such a weight function, we have

$$\Delta_n^{-1/2}(\hat{\gamma}_{D_n} - \gamma) \xrightarrow{\mathscr{L}} N\left[0, \frac{1}{\int_0^1 \left[\frac{\partial}{\partial\gamma}h(t, \gamma)\right]^2 dt}\right] \quad \text{as } \|D_n\| \to 0$$

(4.29)

where $\Delta_n = t_{in} - t_{i-1,n}$ for $1 \le i \le k_n = \Delta_n^{-1}$. In particular if $\Delta_n = 1/k_n$ and $k_n = n$, then

$$n^{1/2}(\gamma_{D_n} - \gamma) \xrightarrow{\mathscr{L}} N\left[\frac{1}{\int_0^1 \left[\frac{\partial}{\partial\gamma}h(t, \gamma)\right]^2 dt}\right]$$

(4.30)

Let us now consider some special cases.
(a) Suppose

$$h(t, \gamma) = e^{-\gamma t}, \quad 0 \le t \le 1$$

(4.31)

Then

$$k_n^{1/2}(\hat{\gamma}_{D_n} - \gamma) \xrightarrow{\mathscr{L}} N\left[0, \frac{1}{\int_0^1 t^2 \, dt}\right] \quad \text{as } \|D_n\| \to 0$$

and hence

$$k_n^{1/2}(\hat{\gamma}_{D_n} - \gamma) \xrightarrow{\mathscr{L}} N(0, 3) \quad \text{as } \|D_n\| \to 0$$

(4.32)

(b) Suppose

$$h(t, \gamma) = e^{-\gamma t}, \quad 0 \le t \le 1$$

Then

$$\frac{\partial}{\partial\gamma}h(t, \gamma) = -te^{-\gamma t}$$

and

$$\int_0^1 \left[\frac{\partial}{\partial\gamma}h(t, \gamma)\right]^2 dt = \int_0^1 t^2 e^{-2\gamma t} \, dt = \psi(\gamma) \quad \text{(say)}$$

(4.33)

Then $\psi(\gamma)$ can be explicitly computed and

$$k_n^{1/2}(\hat{\gamma}_{D_n} - \gamma) \xrightarrow{\mathscr{L}} N\left(0, \frac{1}{\psi(\gamma)}\right) \quad \text{as } \|D_n\| \to 0$$

(4.34)

(c) Suppose

$$h(t, \gamma) = L + (U - L)e^{-e^{A - \gamma t}} \qquad 0 \le t \le 1 \qquad (4.35)$$

where L, U and A are known and γ is unknown parameter. Then

$$\frac{\partial h(t, \gamma)}{\partial \gamma} = (U - L)e^{-e^{A - \gamma t}}(-e^{(A - \gamma t)})(-t)$$

and

$$\int_0^1 \left[\frac{\partial}{\partial \gamma} h(t, \gamma) \right]^2 dt = (U - L)^2 \int_0^1 (e^{-e^{A - \gamma t}} e^{(A - \gamma t)})^2 t^2 \, dt$$

and

$$n^{1/2}(\hat{\gamma}_{D_n} - \gamma) \xrightarrow{\mathscr{L}} N \left[0, \frac{1}{(U - L)^2 \int_0^1 (e^{-e^{A - \gamma t}} e^{(A - \gamma t)})^2 t^2 \, dt} \right] \qquad (4.36)$$

The case of growth curves where $h(t, \gamma)$ is as defined by (c) was discussed in Pasternak and Shohoji (1976) for fitting the model to adolescent standing height growth data when all the parameters L, U, A, are unknown. The following discussion is from Pasternak and Shohoji (1976).

Note that $h(t, \gamma)$ may be modeled as the standing height at age t of an individual. Then $h(t, \gamma) = L$ represents the standing height attained by the individual before the start of adolescent growth cycle and $h(t, \gamma) = U$ represents the height to be attained by the end of the growth cycle. In other words $(U - L)$ is the total contribution to the standing height during the complete growth cycle. Furthermore, if

$$W(t, \gamma) = \frac{h(t, \gamma) - L}{U - L} = e^{-e^{(A - \gamma t)}} \qquad (4.37)$$

then

$$\frac{d \log W(t, \gamma)}{dt} = -\gamma \log W(t, \gamma) \qquad (4.38)$$

Hence $W(t, \gamma)$ represents the proportion of growth attained by age t and

$$e^{-\gamma} = \frac{\log W(t + 1, \gamma)}{\log W(t, \gamma)} \qquad (4.39)$$

can be interpreted as the index of individual growth power. Hence estimation of γ is an important problem in the study of the growth model discussed here.

Example 4.5. Suppose we model the growth by the relation

$$Y_{t_{i,n}} = \beta Y_{t_{i-1,n}} + \epsilon_{t_{i-1,n}} \qquad 1 \le i \le k_n \qquad (4.40)$$

where $Y_{t_{0,n}} = Y_0$ and $\epsilon_{t_{i,n}}$ are independent $N(0, \sigma^2)$. For instance, if we are measuring the height of an individual, the height at time $t_{i,n}$ will depend naturally on the height at time $t_{t-1,n}$ taken earlier. It is easy to see that the weighted likelihood function corresponding to weight function $\lambda(.)$ is proportional to

$$L_n(\beta) = \prod_{i=1}^{k_n} \exp\left[\left\{-\frac{1}{2\sigma^2}(Y_{t_{i,n}} - \beta Y_{t_{i-1,n}})^2\right\}\{\lambda(t_{i,n}) - \lambda(t_{i-1,n})\}\right]$$

$$(4.41)$$

Hence

$$\frac{d \log L_n(\beta)}{d\beta} = \frac{1}{\sigma^2} \sum_{i=1}^{k_n}(Y_{t_{i,n}} - \beta Y_{t_{i-1,n}})Y_{t_{i-1,n}}\{\lambda(t_{i,n}) - \lambda(t_{i-1,n})\}$$

and the RHS is equal to zero implies that

$$\hat{\beta}_{D_n} = \frac{\displaystyle\sum_{i=1}^{k_n} Y_{t_{i,n}} Y_{t_{i-1,n}}\{\lambda(t_{i,n}) - \lambda(t_{i-1,n})\}}{\displaystyle\sum_{i=1}^{k_n} Y_{t_{i-1,n}}^2\{\lambda(t_{i,n}) - \lambda(t_{i-1,n})\}} \qquad (4.42)$$

It is easy to check that $\hat{\beta}_{D_n}$ is a weighted maximum likelihood estimator since

$$\frac{d^2 \log L_n(\beta)}{d\beta^2} < 0 \quad \text{at } \beta = \hat{\beta}_{D_n}$$

Note that

$$\hat{\beta}_{D_n} - \beta = \frac{\displaystyle\sum_{i=1}^{k_n} \epsilon_{t_{i-1,n}} Y_{t_{i-1,n}}(\lambda(t_{i,n}) - \lambda(t_{i-1,n}))}{\displaystyle\sum_{i=1}^{k_n} Y_{t_{i-1,n}}^2(\lambda(t_{i,n}) - \lambda(t_{i-1,n}))} \qquad (4.43)$$

Let

$$R_n = \sum_{i=1}^{k_n} \epsilon_{t_{i-1,n}} Y_{t_{i-1,n}} \{\lambda(t_{i,n}) - \lambda(t_{i-1,n})\} \qquad (4.44)$$

Then

$$E(R_n) = 0 \qquad (4.45)$$

since $Y_{t_{i-1,n}}$ is independent of $\epsilon_{t_{i-1,n}}$ and $E(\epsilon_{t_{i-1,n}}) = 0$. Let $\mathcal{F}_{jn}$ denote the σ-algebra generated by $Y_{t_{0,n}}$ and $\epsilon_{t_{i,n}}, 0 \le i \le j$. Observe that

$$E[\epsilon_{t_{j,n}} Y_{t_{j,n}} \mid \mathcal{F}_{j-1,n}] = Y_{t_{j,n}} E[\epsilon_{t_{j,n}} \mid \mathcal{F}_{j-1,n}]$$

since $Y_{t_{j,n}}$ is $\mathcal{F}_{j-1,n}$ measurable from the relation (4.40). But $E[\epsilon_{t_{j,n}} \mid \mathcal{F}_{j-1,n}] = 0$ since $\epsilon_{t_{j,n}}$ is independent of $\mathcal{F}_{j-1,n}$ and $E(\epsilon_{t_{j,n}}) = 0$. Hence R_n is a sum of martingale differences and

$$\begin{aligned}
\mathrm{Var}(R_n) &= \sum_{i=1}^{k_n} \mathrm{Var}[\epsilon_{t_{i-1,n}} Y_{t_{i-1,n}}][\lambda(t_{i,n}) - \lambda(t_{i-1,n})]^2 \\
&= \sum_{i=1}^{k_n} E[\epsilon_{t_{i-1,n}} Y_{t_{i-1,n}}]^2 [\lambda(t_{i,n}) - \lambda(t_{i-1,n})]^2 \\
&= \sum_{i=1}^{k_n} \sigma^2 E[Y_{t_{i-1,n}}]^2 [\lambda(t_{i,n}) - \lambda(t_{i-1,n})]^2 \qquad (4.46)
\end{aligned}$$

Let

$$T_n = \sum_{i=1}^{k_n} Y_{t_{i-1,n}}^2 [\lambda(t_{i,n}) - \lambda(t_{i-1,n})] \qquad (4.47)$$

Then it follows that

$$T_n(\hat{\beta}_{D_n} - \beta) = R_n \qquad (4.48)$$

One can give sufficient conditions for asymptotic normality of R_n following results in Hall and Heyde (1980) but we will not discuss the details here.

Let us suppose that $\lambda(.)$ is continuously differentiable and $t_{i,n} - t_{i-1,n} \simeq \Delta_n^{-1}$ for $1 \le i \le k_n$ as in Example 4.4. Then

$$\begin{aligned}
\mathrm{Var}(R_n) &\simeq \sigma^2 \Delta_n \sum_{i=1}^{k_n} E[Y_{t_{i-1,n}}]^2 [\lambda'(t_{i-1,n})]^2 (t_{i,n} - t_{i-1,n}) \\
&\simeq \sigma^2 \Delta_n \int_0^1 E[Y_t]^2 [\lambda'(t)]^2 \, dt \qquad (4.49)
\end{aligned}$$

and

$$T_n \Delta_n^{-1/2}(\hat{\beta}_{Dn} - \beta) = \Delta_n^{-1/2} R_n \qquad (4.50)$$

and $\Delta_n^{-1/2} R_n$ is asymptotically normal with mean 0 and variance $\sigma^2 \int_0^1 E[Y_t]^2 [\lambda'(t)]^2 \, dt$. Hence

$$T_n \Delta_n^{-1/2}(\hat{\beta}_{Dn} - \beta) \xrightarrow{\mathscr{L}} N\left(0, \sigma^2 \int_0^1 E[Y_t^2](\lambda'(t))^2 \, dt\right) \qquad (4.51)$$

as $n \to \infty$. Note that T_n will behave asymptotically as

$$\int_0^1 Y_t^2 \lambda'(t) \, dt \qquad (4.52)$$

as $n \to \infty$ and $\hat{\beta}_{Dn}$ is asymptotically normal provided T_n is asymptotically a degenerate random variable.

Example 4.6.   Suppose we model the growth by the relation

$$Y_{t_{i,n}} = \beta t_{i-1,n} Y_{t_{i-1,n}} + \epsilon_{t_{i-1,n}} \qquad 1 \le i \le k_n \qquad (4.53)$$

where $Y_{t_{0,n}} = Y_0$ and $\epsilon_{t_{i,n}}$ are independent $N(0, \sigma^2)$. This model takes into account the time of previous observation unlike the one discussed in the previous example. It is easy to check that the weighted maximum likelihood estimator corresponding to weight function $\lambda(.)$ is

$$\hat{\beta}_{Dn} = \frac{\displaystyle\sum_{i=1}^{k_n} Y_{t_{i,n}} t_{i-1,n} Y_{t_{i-1,n}}(\lambda(t_{in}) - \lambda(t_{i-1,n}))}{\displaystyle\sum_{i=1}^{k_n} t_{i-1,n}^2 Y_{t_{i-1,n}}^2 (\lambda(t_{i,n}) - \lambda(t_{i-1,n}))} \qquad (4.54)$$

and

$$\hat{\beta}_{Dn} - \beta = \frac{\displaystyle\sum_{i=1}^{k_n} \epsilon_{t_{i-1,n}} Y_{t_{i-1,n}} t_{i-1,n}(\lambda(t_{i,n}) - \lambda(t_{i-1,n}))}{\displaystyle\sum_{i=1}^{k_n} t_{i-1,n}^2 Y_{t_{i-1,n}}^2 (\lambda(t_{i,n}) - \lambda(t_{i-1,n}))}$$

Calculations similar to those given in Example 4.5 show that

$$T_n^*(\hat{\beta}_{Dn} - \beta) = R_n^* \qquad (4.55)$$

where

$$T_n^* = \sum_{i=1}^{k_n} t_{i-1,n}^2 Y_{t_{i-1,n}}^2 (\lambda(t_{i,n}) - \lambda(t_{i-1,n})) \tag{4.56}$$

and

$$R_n^* = \sum_{i=1}^{k_n} \epsilon_{t_{i-1,n}} Y_{t_{i-1,n}} t_{i-1,n} (\lambda(t_{i,n}) - \lambda(t_{i-1,n})) \tag{4.57}$$

Suppose $\lambda(.)$ is continuously differentiable and $t_{i,n} - t_{i-1,n} \simeq \Delta_n^{-1}$ for $1 \le i \le k_n$. Then

$$\text{Var}(R_n^*) \simeq \sigma^2 \Delta_n \int_0^1 t^2 E(Y_t^2)[\lambda'(t)]^2 \, dt \tag{4.58}$$

and

$$T_n^* \Delta_n^{-1/2}(\hat{\beta}_{D_n} - \beta) \xrightarrow{\mathcal{L}} N\left(0, \sigma^2 \int_0^1 t^2 E(Y_t^2)[\lambda'(t)]^2 \, dt\right) \tag{4.59}$$

as $n \to \infty$. Furthermore T_n^* behaves asymptotically as

$$\int_0^1 t^2 Y_t^2 \lambda'(t) \, dt \tag{4.60}$$

as $n \to \infty$.

5 REMARKS

The definition of weighted likelihood involves the choice of weight function $\lambda(.)$ from the class Λ of all nondecreasing functions defined on $[0, 1]$. One way of choosing λ is to select $\lambda = \lambda^*$ from Λ so that the asymptotic variance of WMLE corresponding to λ^* is the smallest compared to the asymptotic variances of WMLEs arising from all other weight functions in the class Λ, assuming, of course, that the limiting distribution of WMLE is normal for all weight functions λ in Λ. This involves explicit computation of the asymptotic variance in a closed form as a functional on Λ. This can be done in specific problems, for instance, see Example 4.4. It was shown that the optimum choice of λ is $\lambda(t) \equiv t$ in the subclass Λ^* consisting of all nondecreasing differentiable function on $[0, 1]$ with $\lambda(0) = 0$ and $\lambda(1) = 1$. In general, the choice of $\lambda(.)$ might also depend on information available on the problem as well as other practical considerations. In order to get an idea of the "relative efficiency" of WMLE with respect to the usual MLE, let us look at Example 4.1 again. Let $\beta_{D_n}^*$ denote the ordinary MLE of β estimated from observations $\{y_{t_{in}}, 1 \le i \le k_n\}$ and $\hat{\beta}_{D_n}$ denote the WMLE corresponding to a weight function $\lambda(.)$ and observation scheme D_n. It is easy to see that $\beta_{D_n}^*$ and β_{D_n} are unbiased and have exact normal distribution with variances given by

$$\mathrm{Var}(\beta_{D_n}^*) = \frac{1}{\displaystyle\sum_{i=1}^{k_n} t_{i-1,n}^2}$$

and

$$\mathrm{Var}(\hat{\beta}_{D_n}) = \frac{\displaystyle\sum_{i=1}^{k_n} t_{i-1,n}^2 (\lambda(t_{i,n}) - \lambda(t_{i-1,n}))^2}{\left[\displaystyle\sum_{i=1}^{k_n} t_{i-1,n}^2 (\lambda(t_{i,n}) - \lambda(t_{i-1,n})) \right]^2}$$

Hence

$$\frac{\mathrm{Var}(\hat{\beta}_{D_n})}{\mathrm{Var}(\beta_{D_n}^*)} = \frac{\left[\displaystyle\sum_{i=1}^{k_n} t_{i-1,n}^2 (\lambda(t_{i,n}) - \lambda(t_{i-1,n}))^2 \right] \displaystyle\sum_{i=1}^{k_n} t_{i-1,n}^2}{\left[\displaystyle\sum_{i=1}^{k_n} t_{i-1,n}^2 (\lambda(t_{i,n}) - \lambda(t_{i-1,n})) \right]^2}$$

If $\lambda(t) \equiv t$, then

$$\frac{\mathrm{Var}(\hat{\beta}_{D_n})}{\mathrm{Var}(\beta_{D_n}^*)} \cong 1$$

and, in general, if $\lambda(.)$ is a smooth differentiable function on $[0, 1]$, then

$$\varlimsup_{n \to \infty} \frac{\mathrm{Var}(\hat{\beta}_{D_n})}{\mathrm{Var}(\beta_{D_n}^*)} \le \varlimsup_{n \to \infty} \frac{\|D_n\| \displaystyle\sum_{i=1}^{k_n} t_{i-1,n}^2}{\displaystyle\int_0^1 t^2 \lambda'(t)\, dt}$$

Suppose the sequence of subdivisions $\{D_n\}$ is such that

$$\varlimsup_n \|D_n\| \sum_{i=1}^{k_n} t_{i-1,n}^2 \le c < \infty$$

(For equal subdivision of $[0, 1]$, it is easy to check that $c = 1/3$). Then

$$\varlimsup_{n} \frac{\text{Var}(\hat{\beta}_{D_n})}{\text{Var}(\beta^*_{D_n})} \leq \frac{c}{\int_0^1 t^2 \lambda'(t)\, dt}$$

and the WMLE $\hat{\beta}_{D_n}$ corresponding to the weight function $\lambda(.)$ is more efficient than the ordinary MLE $\beta^*_{D_n}$ asymptotically provided

$$\int_0^1 t^2 \lambda'(t)\, dt > c$$

Further investigation is needed to study whether there is an optimum way to choose $\lambda(.)$ depending on the sequence of subdivisions D_n in the general framework.

Finally, it is worth mentioning that the concepts and results in this paper can be extended to the case when the measurements are made not longitudinally but in spatial framework. The weight function λ here is to be replaced by an appropriate measure on the space of values of t. We will discuss this problem in another publication.

ACKNOWLEDGMENT

The author thanks the referee for his suggestions and comments.

REFERENCES

Basawa, I. V., and Prakasa Rao, B. L. S. (1980). *Statistical Inference for Stochastic Processes.* Academic Press, London.

Basawa, I. V., and Scott, D. J. (1983). *Asymptotic optimal inference for non-ergodic models,* Lecture Notes in Statistics, *17,* Springer-Verlag, New York.

Hall, P., and Heyde, C. (1980). *Martingale Limit Theory and Its Application.* Academic Press, New York.

Pasternak, B. S., and Shohoji, T. (1976). Filling a Gompertz curve to adolescent standing height growth data. *Essays in Probability and Statistics in Honor of Professor Junjiro Ogawa,* 559–577.

Shohoji, T. (1982). On the asymptotic properties of the maximum likelihood estimates in a growth curve. *Technical Report No. 63,* Hiroshima University.

8
Markov Chain Models for Type-Token Relationships

D. J. Daley
Australian National University, Canberra, Australia

J. M. Gani
University of California at Santa Barbara, Santa Barbara, California

D. A. Ratkowsky
Washington State University, Pullman, Washington

This paper is concerned with the fit of some Markov chain models to the type-token relationship in literary texts. The basic model assumes a transition probability $p_{v,v+1} = \gamma(1 - v/M)^{\beta'}$ for the increase from $V_n = v$ to $V_{n+1} = v + 1$ types (distinct words) when an additional token (word) is added to a text of length n. Data from some French authors and *The Bible in Basic English* are analyzed, and two estimation methods are used to obtain $\hat{M}$, $\hat{\gamma}$, $\hat{\beta}$, as well as related parameters for variants of this model. While the fit of the data to the model appears reasonable, chi-square goodness of fit values are large owing to the size of the data sets. The paper derives some useful formulas for the type-token relationship, the simplest of which is

$$EV_n = \left(\frac{1}{\beta}\right) \log(1 + n\gamma\beta)$$

where $\beta'/M \to \beta$ for large β' and M. It is pointed out that the model applies equally well to counts of letters in a text, and an example is given using the Spanish alphabet.

The paper concludes with a mathematical discussion of further refinements to the model.

1 INTRODUCTION

One of the more interesting problems of statistical linguistics is the *type-token* relationship, which is concerned with the number V_n of different words (types) appearing in a text of n words (tokens). This is of particular interest for larger values of n. Clearly, a text always starts with a new (distinct) word so that when $n = 1$, $V_1 = 1$, but as n increases, the number V_n grows more and more slowly. For example, Figure 8.1 shows a logarithmic plot of the points (n, V_n) for the data of Table 8.1 in which the total vocabulary is some finite integer M (here known to be 1000), which is in general some unknown parameter.

There is a large literature of empirical models for the type-token relationship. Among the earliest are those of Yule (1944) and Guiraud (1959), the latter proposing that $V_n/\sqrt{n}$ may be approximately constant. Herdan in a series of books (e.g. Herdan, 1960) culminating in his final monograph (1964) has suggested that $(\log V_n)/(\log n)$ is a constant. These ratios are listed for *The Bible in Basic English*, for which the vocabulary is limited to 1000 types, in Table 8.1, illustrating (as others have) that neither of the proposed relationships holds true in general. For additional references we simply refer the reader to reviews by Gani (1985) and Ratkowsky (1988).

Some theoretically based models have also been considered by Brainerd (1972), Gani (1975), and Sichel (1986), among others. Of these, the one that appears intuitively to reflect the *linguistic* process most closely is the homogeneous Markov chain model in which the transition probability of increasing the number V_n in n tokens to $V_{n+1} = V_n + 1$ in $n + 1$ tokens is

$$p_{v,v+1} \equiv \Pr\{V_{n+1} = v + 1 \mid V_n = v\} = \gamma\left(1 - \frac{v}{M}\right) \quad \gamma > 0 \quad (1.1)$$

Note that when $\gamma = 1$ this is equivalent to selecting at random one of the $M - v$ remaining unused types. This is a simpler form of Brainerd's more general nonhomogeneous Markov chain in which

$$p_{v,v+1} = f(n, v) \quad (1.2)$$

for some general function $f(n, v)$ of both n and v; Brainerd further simplified $f(n, v)$ to a function $g(n)$ dependent on n alone. Gani's (1975) model assumed dependence on v alone as we have already indicated, using a function of the form

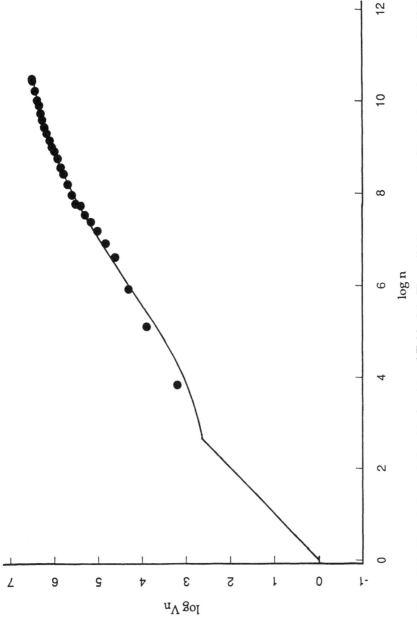

Figure 8.1 Type-token relationship for data of Table 8.1. The solid line represents fitted values from model (3.5) using the parameter estimates in Table 8.6.

Table 8.1   Type-Token Data for *The Bible in Basic English*, Book of Genesis

n	V_n	$V_n/\sqrt{n}$	$\log n$	$\log V_n$	$(\log V_n)/(\log n)$
48	25	3.608	3.871	3.219	0.831
168	50	3.858	5.124	3.912	0.763
380	75	3.847	5.940	4.317	0.727
747	100	3.659	6.616	4.605	0.696
1,020	125	3.914	6.928	4.828	0.697
1,341	150	4.096	7.201	5.011	0.696
1,625	175	4.341	7.393	5.165	0.699
1,899	200	4.590	7.549	5.298	0.702
2,222	225	4.773	7.706	5.416	0.703
2,438	250	5.063	7.799	5.521	0.708
2,850	275	5.151	7.955	5.617	0.706
3,671	300	4.951	8.208	5.704	0.695
4,634	325	4.774	8.441	5.784	0.685
5,368	350	4.777	8.588	5.858	0.682
6,575	375	4.625	8.791	5.927	0.674
7,464	400	4.630	8.918	5.991	0.672
8,303	425	4.664	9.024	6.052	0.671
9,641	450	4.583	9.174	6.109	0.666
11,113	475	4.506	9.316	6.163	0.662
12,882	500	4.405	9.464	6.215	0.657
14,926	525	4.297	9.611	6.263	0.652
17,519	550	4.155	9.771	6.310	0.646
20,571	575	4.009	9.932	6.354	0.640
23,094	600	3.948	10.047	6.397	0.637
28,151	625	3.725	10.245	6.438	0.628
34,184	650	3.516	10.440	6.477	0.620
35,438	665[a]	3.533	10.476	6.500	0.620

[a]This last number is 665, *not* 675, as Genesis consists of 35,438 tokens and
precisely 665 types.

$$h(v) = \gamma \left(1 - \frac{v}{M} \right)$$

Sichel (1986) is more concerned with fitting a probability distribution
to the observed data for V_n and n, proposing the simplified relationship

$$V_n \approx 2(bc)^{-1}\{1 - \exp[-b((1 + cn)^{1/2} - 1)]\} \qquad (1.3)$$

While this model provides a satisfactory fit to his data, the parameters

b and c do not appear to remain constant for a single author as Sichel suggests from his Table 4. For example, Table 8.2 presents least squares estimates of the parameters of this model using individual plays, chapters, or sections of a work or corpus of works by two French authors. The estimates are clearly too variable, compared with their standard errors, for b and c to be considered constant for each author.

The fit of the model with $p_{v,v+1} = h(v)$ to actual data from several French authors likewise proves unsatisfactory. In this paper we propose refinements of the model in order to achieve improved fits. Since experience indicates that, as an author's vocabulary nears its limit M, the probability of producing a new type becomes increasingly smaller, we propose within the Markov chain framework to use the transition probabilities

$$p_{v,v+1} = \gamma \left(1 - \frac{v}{M} \right)^{\beta'} \qquad \gamma > 0 \tag{1.4}$$

where β' is larger than 1, so that $p_{v,v+1} < \gamma(1 - v/M)$, with the inequality becoming more pronounced as v approaches M.

We develop some consequences of this model, indicating a further change in which we replace the right-hand side by

$$\gamma e^{-v\beta} \tag{1.5}$$

Table 8.2 Least Squares Estimates and Standard Errors of Parameters b and c in Sichel's (1986) Model (1.3) for Works of Corneille and Racine

Author and work	m	n	V_n	$100(\hat{b} \pm \text{S.E.})$	$100(\hat{c} \pm \text{S.E.})$
Corneille					
Sertorius	20	17,710	1,546	4.85 ± 0.34	1.36 ± 0.04
Sophonisbe	19	16,891	1,425	4.80 ± 0.45	1.49 ± 0.06
18 Tragedies	18	303,353	3,562	3.94 ± 0.21	1.32 ± 0.06
Total theatre	32	532,800	4,606	2.68 ± 0.11	1.46 ± 0.05
Racine					
Phèdre	17	14,417	1,580	3.03 ± 0.30	1.38 ± 0.04
Andromaque	17	15,126	1,230	4.09 ± 0.31	1.96 ± 0.06
Mitridate	17	15,128	1,360	1.61 ± 0.38	2.17 ± 0.11
Tragedies	11	158,899	2,931	1.84 ± 0.38	2.72 ± 0.32

Note: m is the number of chapters, plays, or sections of the work. The least squares criterion was applied to cumulative values of n and V_n obtained by combining, chronologically or sequentially, these individual portions of the work.

where $\beta = \beta'/M$. We then proceed to fit the model to counts based on the works of several French authors (Constant, Racine, Corneille, Giraudoux, and Zola) as well as *The Bible in Basic English* (1968), with its vocabulary of 1000 types. The model also fits counts of letters in the Spanish alphabet, as indicated in the discussion of Tables 8.9 and 8.10. In each case estimates of γ and β, or of γ, β' and M, are obtained, and some discussion is given of the stability of these estimates and of the goodness-of-fit of the model. We begin by outlining some possible models in more detail.

2 SOME POSSIBLE MODELS

Among the assumptions underlying the types and tokens models in the existing literature are the following:

A_0:   There exists a finite (albeit large) vocabulary of, say, M_0 words available for inclusion in *any* piece of writing.

A_a: Each author a has a finite resource vocabulary of size $M_a < M_0$.

B: As a piece of writing grows from a total of n words (=tokens) utilizing a vocabulary of V_n distinct words (i.e., V_n types), the propensity of an author to increase the utilized vocabulary decreases (i.e., with increasing n, it becomes less likely that $V_{n+1} = V_n + 1$ and correspondingly more likely that $V_{n+1} = V_n$); this can be viewed as reflecting (A_a) in part.

The following simple probability model covering (A_a) and (B), while not very realistic, is of use; it also motivates subsequent developments. Assume that a piece of writing consists of words drawn *at random* and *uniformly* from the resource vocabulary of size M_a. These two assumptions imply that

$$\Pr\{V_{n+1} = v + 1 \mid V_n = v, V_{n-1}, \ldots\} = \Pr\{V_{n+1} = v + 1 \mid V_n = v\}$$

(2.1)

$$\equiv \alpha_v \equiv p_{v,v+1} = \frac{(M_a - v)}{M_a} = 1 - \frac{v}{M_a}$$

(2.1')

= proportion of resource vocabulary not used in first n words.

We can as easily define, for $v = 1,2,\ldots,$

$$N_v \equiv \inf\{n\colon V_n = v\}$$

Then the Markovian property at (2.1) implies that N_ν is the sum of ν independent geometrically distributed random variables $T_1, \ldots, T_\nu$, where for $r = 1, 2, \ldots,$

$$\Pr\{T_\nu = r\} = \alpha_{\nu-1}(1 - \alpha_{\nu-1})^{r-1}$$

Thus,

$$E(N_\nu) = \sum_{i=1}^{\nu} \frac{1}{\alpha_{i-1}} \tag{2.2}$$

$$= \sum_{i=0}^{\nu-1} \frac{M_a}{M_a - i} \approx -M_a \log\left(1 - \frac{\nu}{M_a}\right) \tag{2.2'}$$

In practice, we should not expect the uniformity assumption underlying (2.1') to be met. We indicate in Section 5 some motivation for retaining the Markovian property of $\{V_n\}$ at (2.1) but with the modified definition of the probability α_ν as

$$\alpha_\nu = \gamma\left(1 - \frac{\nu}{M}\right)^{\beta'} \tag{2.3}$$

for some parameters γ and β' to be determined. We continue to regard these parameters as varying between authors, and possibly between different works of each author, so that they should perhaps be indexed by a as a reminder of this dependence: it is notationally simpler to take this for granted. At the same time, we shall indicate why determining M ($\equiv M_a$) in this model can be expected to be difficult, and that it may be better instead to use the parametric form

$$\alpha_\nu = \gamma \exp(-\nu\beta) \qquad \beta \equiv \frac{\beta'}{M} \tag{2.4}$$

In practice, while the Markovian assumption at (2.1) may be a gross simplification, it has the advantage of furnishing an approximation to an error variance term in a piecewise linear version of the model. For this, suppose we are given data $\{V_n\}$ for $n = n_j$, $j = 0, 1, \ldots, J$, with $0 = n_0 < n_1 < \cdots < n_J$. Then, conditional on $V_{n_{j-1}}$, the increment $\Delta_j \equiv V_{n_j} - V_{n_{j-1}}$ is independent of n_{j-1} and $\{V_n: n < n_{j-1}\}$; it is a random variable with distribution determined entirely by $V_{n_{j-1}}$, $\nu_j \equiv n_j - n_{j-1}$ and $\{\alpha_\nu: \nu \geq V_{n_{j-1}}\}$. To a fair degree of approximation, Δ_j is distributed as a binomial random variable with parameters (ν_j, α_ν) where $\nu = V_{n_{j-1}}$, so that Δ_j has approximate mean and variance

$$\nu_j \alpha_\nu \qquad \nu_j \alpha_\nu (1 - \alpha_\nu) \qquad\qquad (2.5)$$

respectively. Granted the "smoothness" of α_ν, a better approximation ensues on replacing α_ν here by an averaged value $\bar{\alpha}_\nu \equiv \alpha_{\nu + \Delta/2}$. This means that we can fit the model based on the Markov property at (2.1) but with α_ν (or $\bar{\alpha}_\nu$) given by (2.4), using the standard likelihood estimation procedure of maximizing

$$\prod_{j=1}^{J} [(2\pi\nu_j \bar{\alpha}_{\nu_{j-1}}(1 - \bar{\alpha}_{\nu_{j-1}}))]^{-1/2} \exp\left[- \frac{(\Delta_j - \nu_j \bar{\alpha}_{\nu_{j-1}})^2}{2\nu_j \bar{\alpha}_{\nu_{j-1}}(1 - \bar{\alpha}_{\nu_{j-1}})} \right] \qquad (2.6)$$

Assuming the model to be reasonable, the sum

$$\sum_{j=1}^{J} \frac{(\Delta_j - \nu_j \bar{\alpha}_{\nu_{j-1}})^2}{\nu_j \bar{\alpha}_{\nu_{j-1}}(1 - \bar{\alpha}_{\nu_{j-1}})} \qquad\qquad (2.7)$$

is distributed as a chi-square random variable on $J - \nu$ degrees of freedom where ν is the number of parameters estimated. We shall use this property later when goodness-of-fit tests are considered.

3   DATA SETS AND APPROACHES TO PARAMETER ESTIMATION

Six data sets on the prose or theatre works of the French authors Constant, Racine, Corneille, Giraudoux, Proust, and Zola have been compiled in the form of cumulative values of (n, V_n) by various workers in quantitative linguistics (Allen, 1984; Bernet, 1983; Brunet, 1978, 1983, 1985; Dugast, 1979). For example, in an individual work such as Constant's *Adolphe*, the raw data come from each of its ten chapters. Cumulative values of the number of tokens n and the number of types V_n were obtained by noting the value of n at which each new type appears. The data were presented (Allen, 1984) only as totals for each of the ten chapters as in Table 8.3. A graph of (n, V_n) would result in a set of points similar to that in Figure 8.1.

For an entire corpus, such as all of Corneille's works for the theatre, Dugast (1979) reported cumulative values of (n, V_n) as the totals of each of the 32 plays taken chronologically. Similar decisions about the points in the corpus at which to report cumulative values of (n, V_n) were made by the various other workers referenced and whose tabulated data are used in this study.

To these six data sets we have added the further data set shown in

Table 8.3 Cumulative Type-Token Data for Constant's *Adolphe*

Chapter	Cumulative n	Cumulative V_n	Chapter	Cumulative n	Cumulative V_n
1	2,040	659	6	16,790	2,040
2	5,710	1,260	7	19,830	2,203
3	8,160	1,490	8	22,360	2,330
4	10,990	1,675	9	23,970	2,377
5	14,240	1,905	10	27,660	2,509

Table 8.1 based on the first part of *The Bible in Basic English* (1982) (see also Ogden, 1968), that was compiled using its known vocabulary size $M = 1,000$. We obtained cumulative values of (n, V_n) by employing the same procedure as other workers (see Table 8.1). We now have the choice of two possible ways of obtaining the parameter estimates for the models considered here. One method, which we also used for the French literary texts, is to fit the observed data (n, V_n) to a curve of expected values such as in Figure 8.1, using the method of least squares. The other method takes advantage of the Markovian structure (2.1) of the model, and fits the observed to the predicted increments in types for fixed increments in the number of tokens. Here, after making some smoothness approximations, we used a maximum likelihood approach to estimate the model parameters; since the distributional approximations yield Gaussian distributions, the method is similar to a weighted least squares estimation procedure.

3.1 Least Squares Estimation Procedure

We have indicated that two estimation methods will be used to fit Markov chain models to observations of (n, V_n) where V_n is the number of different types appearing in a text of n tokens. The basic model for $\{V_n, n = 1, 2, \ldots\}$ is a Markov chain with transition probabilities (1.4), subsequently modified to (1.5). The data may be thought of as consisting of observations of V_n at times $n_1, n_2, \ldots$.

Our first estimation method (see the discussion in Section 5) approximates the expected value of V_n by the solution (at integer times) of the equation,

$$D_t(EV_t) = \alpha(EV_t) \qquad t > 0$$

where $D_t = d/dt$, which approximates the exact relation $EV_{n+1} - EV_n = E[\alpha(V_n)]$. Specifically, let α_v be a discrete skeleton of the continuous function $\alpha(v)$, and write

$$n \approx \int_0^{EV_n} \frac{dv}{\alpha(v)} \approx \int_0^{EV_n} \frac{dv}{\gamma\left(1 - \dfrac{v}{M}\right)^{\beta'}}$$

$$= \frac{M}{\gamma(\beta' - 1)}\left[\left(1 - \frac{EV_n}{M}\right)^{-(\beta' - 1)} - 1\right] \qquad (3.1)$$

This approximation gives the expression for EV_n in terms of n as

$$EV_n \approx M[1 - \{1 + \gamma(\beta' - 1)n/M\}^{-1/(\beta' - 1)}] \qquad (3.2)$$

Given data (n, V_n), we have estimated the parameters γ, β', and M by least squares, involving the minimization of the sum of squares of deviations between the observed V_n and the (approximate) expectation EV_n. For the data sets from each of the six French authors, no finite optimum M was obtained, as β' and M both increased without limit in such a way that β'/M approached a constant value (see equations (5.19)–(5.20)). Since

$$\lim_{\beta', M \to \infty} \left(1 - \frac{v}{M}\right)^{\beta'} = e^{-v}(\lim \beta'/M) = e^{-\beta v}$$

for β' and M such that $\beta'/M \to \beta$, the basic model reduces to the following simple expression giving EV_n as a function of n,

$$EV_n = (1/\beta) \log(1 + n\gamma\beta) \qquad (3.3)$$

Numerical results for the estimates of the parameters γ and β for the 7 data sets referenced earlier are given in Table 8.4. The high negative correlation between $\hat{\gamma}$ and the total token count n suggests that γ may be a measure of a quantity decreasing with increasing text or corpus size. Modeling arguments (see also Section 5) suggest that it may be interpreted as the body of "functional" words (i.e., pronouns, prepositions, conjunctions, auxiliary verbs, articles, and interjections) that tend to be independent of context and that are likely to be used by all authors writing in the language concerned (French for all except *The Bible in Basic English* in these data). As n increases, the number of unused functional words decreases, so γ may be thought of as a measure of the ratio of contextual to functional words in the text under consider-

Table 8.4 Parameter Estimates Fitted to Model (3.3) by Least Squares

Author	m	Final n	Final V_n	$\hat{\gamma}$	$10^3\hat{\beta}$
Constant	10	27,660	2,509	0.446	1.062
Racine	11	158,899	2,931	0.203	1.381
Corneille	32	532,800	4,606	0.326	1.121
Giraudoux	23	671,364	15,771	0.211	0.218
Proust	7	1,267,069	18,322	0.135	0.193
Zola	22	2,874,755	19,337	0.091	0.211
Bible in Basic English	27	35,438	665	0.202	5.538

ation. The other parameter β appears to be a direct measure of the size of an author's vocabulary: those who have been traditionally thought to have low vocabulary sizes (e.g., Racine, Corneille and the "authors" of *The Bible in Basic English*) have high values of $\hat{\beta}$, whereas authors of greater vocabulary richness (e.g., Giraudoux, Proust, and Zola) have low but comparable values of $\hat{\beta}$.

Table 8.5 shows both the fitted and observed values for V_n for Zola and Racine. Considering the large size of n, we view these results as illustrating quite good agreement between fitted and observed data. Comparable fits were obtained for the other authors.

Although it is clear from the results that there is reasonably good agreement between the observed and predicted vocabularies, equation (2.3) predicts unrealistically small probabilities when V_n is small, since $\alpha_v \approx \gamma$ for small v and all the estimates of γ are rather less than 1. This is contrary to what is observed (and expected), namely that new types appear with almost every new token for small v. Thus, the parameter estimates in Table 8.4 would be expected to give poor agreement between observed and predicted vocabularies for samples (n, V_n) with small n. A way of overcoming the problem is to modify equation (2.3) so that we fix $\gamma = 1$ for small values of V_n, $v \leq v_c$ (say) for some critical value v_c to be estimated from the data, while γ may take any value for $V_n > v_c$. In other words, we should have

$$\alpha_v = \begin{cases} (1 - v/M)^{\beta'} & v \leq v_c \\ \gamma(1 - v/M)^{\beta'} & v > v_c \end{cases} \tag{3.4}$$

This leads to the following expression for EV_n as a function of n for the case where there is no finite M:

Table 8.5 Comparison of V_n and EV_n Using Equations (3.3) and (3.5) for Works of Racine and Zola

	Zola				Racine		
n	V_n	EV_n (3.3)	EV_n (3.5)	n	V_n	EV_n (3.3)	EV_n (3.5)
68,937	4,606	4,006	4,476	13,828	1,225	1,147	1,262
171,285	6,590	6,912	6,991	27,735	1,596	1,572	1,553
293,228	8,863	8,976	8,921	42,821	1,799	1,857	1,800
402,297	10,362	10,275	10,181	58,252	2,082	2,065	2,005
516,829	11,623	11,340	11,232	71,513	2,182	2,205	2,155
635,356	12,146	12,236	12,127	86,848	2,284	2,340	2,306
757,064	12,934	13,007	12,906	101,992	2,355	2,452	2,437
888,315	13,615	13,719	13,628	117,810	2,448	2,553	2,559
1,053,332	14,655	14,484	14,410	132,225	2,616	2,634	2,659
1,159,993	14,924	14,920	14,857	143,394	2,772	2,692	2,731
1,305,598	15,340	15,457	15,409	158,899	2,931	2,764	2,823
1,445,575	15,664	15,921	15,888				
1,599,066	16,078	16,383	16,366				
1,722,694	16,401	16,725	16,720				
1,895,104	16,977	17,163	17,175				
2,033,405	17,421	17,488	17,512				
2,204,897	17,904	17,862	17,901				
2,275,090	18,151	18,007	18,052				
2,408,571	18,382	18,271	18,327				
2,559,404	18,760	18,552	18,621				
2,754,744	19,127	18,894	18,977				
2,874,755	19,337	19,092	19,184				

$$EV_n = \begin{cases} (1/\beta) \log(1 + n\beta) & v \le v_c \\ (1/\beta) \log[\gamma(1 + n\beta) + (1 - \gamma)e^{\beta v_c}] & v > v_c \end{cases} \quad (3.5)$$

The three parameters of (3.5) were estimated by least squares for the data sets described at the outset, with improvements in the goodness-of-fit occurring at the lowest value of V_n for each data set and sometimes at the highest values also. Table 8.6 presents the values of these estimates. We observe fair agreement between estimates of β from equations (3.3) and (3.5). The results reinforce the interpretation of β as a measure of an author's rate of vocabulary usage with lower values of β consistent with a larger vocabulary. The estimates of γ are fairly consistent between equations (3.3) and (3.5), except for Racine where the value from (3.5) is much lower. It should be recognized that some parameters, as for example v_c, may be doing little more than acting as curve-fitting parameters lacking contextual interpretation; in the case of γ the modeling considerations of Section 5 point to its representing the relative frequency of nonfunctional words.

Table 8.6 Parameter Estimates Fitted to Model (3.5) by Least Squares

Author	$\hat{v}_c$	$\hat{\gamma}$	$10^3\hat{\beta}$
Constant	226	0.389	0.984
Racine	915	0.077	0.920
Corneille	300	0.314	1.111
Giraudoux	645	0.194	0.210
Proust	2,778	0.108	0.178
Zola	1,689	0.078	0.200
Bible in Basic English	14	0.194	5.437

Table 8.5 lists predicted vocabularies for Zola and Racine using equation (3.5). These predicted values, compared with those obtained from equation (3.3) and listed in the same table, show the improvement in goodness-of-fit at the smallest and largest values of V_n. This improvement has been achieved at the cost of introducing an extra parameter v_c into the model. Although there is a discontinuity at $v = v_c$ in the transition probabilities given by (3.4), there is none in the resulting type-token relationship given by equation (3.5) for the expectation of V_n. In Figure 8.1, the solid line displays fitted values of V_n against n, on a logarithmic scale. This line is based on equation (3.5) using the parameter estimates in Table 8.6 for *The Bible in Basic English*, and shows a satisfactory fit. The basic models (3.3) and (3.5) when no finite optimum M exists appear to be adequate for the type-token relationship envisaged here. The algebraic expressions for (3.3) and (3.5) are no more complex in form than Sichel's (1986) two parameter model (1.3).

3.2 Maximum Likelihood Estimation Using Intervals Between Successive Types

The second procedure for estimating the parameters in the model relies on data on the intervals between successive types. Using the procedure described in Section 2, estimates of the parameters γ and β for the data sets considered in this study were obtained by maximizing the likelihood function (2.6). For the case of finite M, α_v is given by (2.3); otherwise, α_v is given by (2.4). When the method was applied to the data sets for each of the French authors and for the data from *The Bible in Basic English* in Table 8.1, a finite optimum M was not found, as β' and M both increased in such a manner that β'/M approaches a

Table 8.7 The Dependence Between β' and M Illustrated by Fitting the Maximum Likelihood Model Based on (2.6) and (2.3) to *The Bible in Basic English* Data

M	$\hat{\beta}'$	$\hat{\gamma}$	$10^3[\hat{\beta} \equiv \hat{\beta}'/M]$	Log-likelihood
700	1.4272	0.115	2.039	−181.1
800	2.2533	0.143	2.817	−144.5
900	2.9333	0.158	3.259	−133.7
1,000	3.5710	0.167	3.571	−128.5
1,500	6.5804	0.191	4.387	−120.3
2,000	9.5075	0.201	4.754	‹ −118.2
4,000	21.0866	0.214	5.272	−116.3
10,000	55.6816	0.221	5.568	−115.6
∞		0.226	5.761	−115.2

constant value β (see equations (5.19) and (5.20)). The results in Table 8.7 illustrate the dependence between β' and M for the data from *The Bible in Basic English* and show the log-likelihood declining monotonically as M increases, despite the fact that M is known to be 1,000 for that work.

Using the definition of α_v given by equation (2.4) for the case of no finite optimum M, results are shown in Table 8.8 for the data sets considered here, along with the chi-square measures of goodness-of-fit given by (2.7). All of the chi-square values are sufficiently large to reject the model as providing an adequate fit (with its attendant assumptions) to the data. However, as argued in Section 4, small local changes in subject matter of a literary work or works may lead to substantial deviations from the constant parameter values that the model assumes.

Table 8.8 Parameter Estimates and Goodness-of-Fit (2.7) Using the Maximum Likelihood Approach (2.6) for the Model (2.4)

Author	m	n	$\hat{\gamma}$	$10^3\hat{\beta}$	χ^2
Constant	10	27,660	0.447	1.058	28.8
Racine	11	158,899	0.163	1.167	381.8
Corneille	32	532,800	0.280	1.064	584.8
Giraudoux	23	671,364	0.210	0.218	619.8
Proust	7	1,267,069	0.116	0.183	69.3
Zola	22	2,874,755	0.094	0.210	1111.1
Bible in Basic English	27	35,438	0.226	5.761	96.8

We note further that the methodology applied here to words in a text may be equally applied to the letters of an alphabet. In that case, the "tokens" are the total numbers of letters, and the "types" are the distinct letters of the alphabet. For the English language, $M = 26$, but one can augment this by regarding punctuation symbols (including a space) and nonstandard characters as distinct types also. Texts in any language should be equally suitable; for example, we obtained a type-token data set based on counts of letters from a portion of Spanish text (see Table 8.9).

When the maximum-likelihood approach based on equation (2.6) was used with the definition for α_v given by equation (2.3), estimates of γ were obtained that were close to 1.0. As noted in Section 1, $\gamma = 1$ is equivalent to selecting an unused type at random. Since the arrangement of letters in a text is more likely to approach randomness than the arrangement of words in the same text, this result is not unexpected. Reestimating the parameter β' for the data in Table 8.9, with α_v given by the model

$$\alpha_v = \left(1 - \frac{v}{M}\right)^{\beta'} \tag{3.6}$$

Table 8.9 Type-Token Data from a Spanish-Language Text Based on the Letters in the Alphabet (with $\tilde{N}$ and LL considered to be distinct from N and L, respectively)

n	V_n	n	V_n
1	1	29	14
2	2	36	15
3	3	43	16
5	4	45	17
6	5	47	18
8	6	54	19
9	7	71	20
10	8	74	21
11	9	121	22
15	10	237	23
18	11	267	24
22	12	309	25
23	13		

Table 8.10 Estimates of β' Using the Maximum
Likelihood Approach (2.6) and χ^2 Goodness-of-
Fit (2.7), Data of Table 8.9, for Different Values
of M, Model (3.6)

M	$\hat{\beta}'$	χ^2 (23 d.f.)
25	1.433	23.62
26	1.755	20.50
27	1.970	20.23
28	2.156	20.38
29	2.328	20.64
30	2.492	20.94

instead of by (2.3), produced the results presented in Table 8.10. The χ^2 value from (2.7) indicates a good fit, with the optimum M being very close to the correct value of 28 (since the Spanish alphabet consists of the 26 letters $A, \ldots, Z$, as well as $\tilde{N}$ and LL, these last two being distinct from N and L, respectively).

4 STABILITY OF PARAMETER ESTIMATES AND GOODNESS-OF-FIT TEST

One way of assessing the adequacy of fit of a model is to examine the extent of variation of the parameter estimates obtained from differing methods of fitting that model. Thus, we can compare the estimates of γ and β in Table 8.4, where the least-squares criterion was employed, with the estimates of γ and β in Table 8.8, that were obtained using the maximum likelihood approach. The agreement between the two sets of estimates is generally very high, the biggest discrepancies, of the order of 20%, occurring for the works of Racine.

A second basic way of assessing the adequacy of fit of a model is to consider the stability or otherwise of the estimated values of parameters as a data set evolves. In the present case, we have done this for each author by estimating γ and β as the number of tokens n increases. Table 8.11 shows the estimates obtained for the seven (sets of) texts considered here for about the first one-third to one-half of the text using the maximum likelihood approach. These can be compared with the results in Table 8.8 obtained for the whole text using the same approach. There are two noticeable effects of this comparison. First,

Table 8.11 Parameter Estimates using the Maximum Likelihood Approach (2.6) for the Model (2.4) Fitted to Part of Various Works

Author	m	n	$\hat{\gamma}$	$10^3\hat{\beta}$
Constant	4	10,990	0.491	1.207
Racine	4	58,252	0.203	1.383
Corneille	10	156,407	0.274	1.055
Giraudoux	8	332,979	0.201	0.208
Proust	3	631,293	0.118	0.187
Zola	7	757,064	0.102	0.225
The Bible in Basic English	19	11,113	0.227	5.782

relative to their values using the full text, the estimates of γ and β from part or the text range between about 20% higher (Racine) and 5% lower (Giraudoux). Second, the changes in the estimates of γ and β for any author are in the same direction, that is, the changes are positively correlated. Granted the size of the underlying lengths of text, we view both the relative stability of the estimates in comparison with the spread for different authors, and the positive correlation of the changes for each author, as indicating that the interpretations of γ and β as measures of the ratio of contextual to functional words and of richness of vocabulary for an author, respectively, have reasonable empirical foundation as well as some basis in terms of the underlying model of Section 5.

The part of this conclusion relating to an empirical basis for the parameters as measures that characterize an author's writing is independent of any stochastic modeling: given the stability and systematic movement of any changes, the conclusion holds for any such function independently of the rationale for considering its use. That it is relevant to model the process follows, first because it suggests the form of the function to be fitted and thereby gives a mechanism to assist in interpreting the parameters, and second because it also provides some kind of goodness-of-fit test as we now indicate.

According to the model, $\{V_n\}$ is a Markov chain of a "purely evolutionary" type, by which we mean that every state of the process is such that the process either enters it once (and stays there for a geometrically or exponentially distributed length of time), or never enters it. In fact, since in this particular case the states are visited in the same order as a random walk proceeding upwards through $1, 2, \ldots$, the estimates given in (2.6) and (2.7) are available. For the seven data sets considered here, the goodness-of-fit, based on (2.7), is given as the last column of Table 8.8. Although these χ^2 values are large, the individual increments Δ_j with mean and variance at (2.5) relate to large values of the parameters ν_j; thus, small perturbations from the (strong) assumption of conditional independence that lead to the variance estimates there can yield significantly different values of those estimates. In terms of the model, local trends in the mean rates $1/\alpha_\nu$ can arise as reflecting changes of context from chapter to chapter or book to book, and these small changes would lead to large deviations from the constant parameters γ and β assumed in the model.

5 FURTHER REMARKS ON MODELING

Recall that the model at (2.1) with α_ν as specified there can be viewed as the embedded jump process of a continuous time Markov chain consisting of the superposition of $M \equiv M_a$ independent and identically distributed Poisson processes with rate λ (say), each of which starts at a common time origin, $t = 0$ (say). Observe that the process $N(t)$ that counts the number of processes in which there has been at least one point by time t is Markovian. It is also relevant for the following to consider the Markov counting process $X(t)$ consisting of the total number of points in the interval $(0, t)$. Define the random times

$$t_n = \inf\{t: X(t) = n\} \qquad t'_n = \inf\{t: N(t) = n\} \qquad (5.1)$$

Then the process $\{V_n\}$ satisfies $V_n = N(t_n)$, while the number of words between the appearance of the νth and $(\nu + 1)$th new words is the difference $T_{\nu+1} \equiv X(t'_{\nu+1}) - X(t'_\nu)$.

It is now a simple matter to change the rate of occurrence of the jth word from λ to $\lambda_j \equiv \lambda_j(a)$ for author a, with $j = 1, \ldots, M \equiv M_a$, so that the relative frequencies of the different words equal λ_j/Λ_0 where $\Lambda_0 = \Sigma_{j=1}^M \lambda_j$. Given text from author a of (time) length t, the frequency of occurrence $X_j(t)$ say of word j is Poisson distributed with mean $\lambda_j t$, the total number of words

$$X(t) = \sum_{j=1}^M X_j(t) \qquad (5.2)$$

is also Poisson distributed with mean $\Lambda_0 t$, while the joint distribution of $\{X_j(t_n)\}$ of the numbers of words that have appeared when the total number equals n has the multinomial distribution with parameters $(n; \{\lambda_j/\Lambda_0\})$. By introducing indicator variables

$$I_j(t) \equiv I_{\{X_j(t)>0\}} \tag{5.3}$$

we can describe the number of different words $N(t)$ and their rate of appearance by

$$N(t) = \sum_{j=1}^{M} I_j(t) \quad \text{and} \quad \Lambda(t) \equiv \sum_{j=1}^{M} \lambda_j[1 - I_j(t)] \tag{5.4}$$

respectively (observe that Λ_0 as introduced earlier equals the initial value $\Lambda(0)$ of the process $\Lambda(\cdot)$). Conditional on $\{I_j(t_n)\}$, which implies that $V_n = N(t_n)$ is given,

$$\Pr\{V_{n+1} = i + 1 \mid V_n = i, \{I_j(t_n)\}\} = \Lambda(t_n)/\Lambda_0 \tag{5.5}$$

thus, to the extent that $\Lambda(t_n)$ varies with $\{I_j(t_n)\}$, so the relationship at (2.1) may be approximately independent or otherwise of $N(t_n)$. For convenience, we record some expectations and inequalities, writing $D_t \equiv d/dt$ in (5.7), namely

$$EN(t) = \sum_{j=1}^{M} [1 - \exp(-\lambda_j t)] \tag{5.6}$$

$$\leq M[1 - \exp(-\Lambda_0 t/M)] = M[1 - \exp(-EX(t)/M)] \tag{5.6'}$$

$$D_t EN(t) = E\Lambda(t) = \sum_{j=1}^{M} \lambda_j \exp(-\lambda_j t) \tag{5.7}$$

$$\leq \Lambda_0 \exp(-\Lambda_0 t/M) = \Lambda_0 \exp(-EX(t)/M) \tag{5.7'}$$

$$\text{var } \Lambda(t) = \sum_{j=1}^{M} \lambda_j^2 e^{-\lambda_j t}(1 - e^{-\lambda_j t}) \tag{5.8}$$

$$\leq \Lambda_0 \sup_j \lambda_j e^{-\lambda_j t} \tag{5.8'}$$

and this last expression decreases with increasing t. In the case of equality of the rates λ_j these expressions simplify considerably, as used in (2.1') and (2.2') for example.

For application to the data of the kind that have been collected, we should in fact replace our expressions here by the more complicated relations that follow from using the multinomial distribution, that is, in place of t we should have t_n and so on. In practice the differences are negligible in comparison with the variable nature of the quantities in the sets $\{\lambda_j\}$.

Proceeding on this basis, start by replacing t in (5.6′) by t_n, so that $X(t_n) = n$ and $N(t_n) = V_n$. We then have

$$E(V_n) \le M(1 - e^{-n/M}) \tag{5.9}$$

whereas approximate equality holds for constant λ. By introducing a convenient parameter $\beta'(n)$ say, with $\beta'(n) > 1$, we can certainly adapt (5.9) into the equation

$$E(V_n) = M(1 - e^{-n/M})^{\beta'(n)} \tag{5.10}$$

This can be linked with (2.3) and (2.4) as follows. Embed V_n into a continuous time Markov chain $V(\cdot)$ whose Q-matrix of transition rates has diagonal elements $-q_i$ and $q_{i,i+1} = \alpha_i = q_i$ as the only off-diagonal nonzero elements in each row. Then

$$D_t E(V(t) \mid V(t - 0) = v) = \alpha_v \tag{5.11}$$

Take expectations over $V(t - 0)$, assume that differentiation and expectation can be interchanged, and that

$$E(\alpha_{V(t-0)}) \approx \alpha(EV(t - 0)) = \alpha(EV(t)) \tag{5.12}$$

using both the continuous function $\alpha(\cdot)$ of which $\alpha_{\cdot}$ is a skeleton and the continuity in t of $EV(t)$. Then an approximate differential equation for $v(t) \equiv EV(t)$ ensues, namely

$$D_t v(t) = \alpha(v(t)) \tag{5.13}$$

The case that $\alpha(v) = \gamma e^{-v\beta}$ yields $D_t v(t) = \gamma e^{-\beta v(t)}$, with solution (for $0 < u < t < \infty$)

$$
\begin{aligned}
v(t) &= \beta^{-1} \log(e^{\beta v(u)} + \gamma\beta(t - u)) \\
&= v(u) + \beta^{-1} \log\{1 + \gamma\beta(t - u)e^{-\beta v(u)}\}
\end{aligned} \tag{5.14}
$$

This then gives for the particular case of the increments Δ_j,

$$E(\Delta_j \mid V_{n_{j-1}} = v) = \beta^{-1} \log(1 + \gamma\beta v_j e^{-\beta v}) \tag{5.15}$$

and $\mathrm{var}(\Delta_j) \approx E(\Delta_j)$ as a first approximation provided γ is somewhat smaller than 1. In case the relation (2.3) is considered preferable, the basic comparable formula is

$$
\begin{aligned}
v(t) = M - (M - v(u)) \\
\times [1 + M\gamma(\beta' - 1)(t - u)(1 - v(u)/M)^{\beta'-1}]^{-1/(\beta'-1)} \tag{5.14′}
\end{aligned}
$$

It is a valid observation based on (5.11) or (5.12) that if instead of a model based on $\{\alpha_v\}$ we are simply given a fitted curve for $v(t)$ (see the specific form at (5.14)) then differentiation yields (5.11) or (5.12) and thus the sequence $\{\alpha_v\}$ needed for the Markov chain model that starts from (2.1) can be deduced in principle.

In Gani and Saunders (1976) consideration was given effectively to using rates $\{\lambda_j\}$ that can take just one of two values, corresponding to a resource vocabulary consisting of words of two subvocabularies, the smaller one being comprised of "functional" words (pronouns, prepositions, conjunctions, auxiliary verbs, articles and interjections), while the other covers "context-dependent" words (nouns, verbs, adjectives, and adverbs) that comprise the vast pool of words that are used in describing both gross and subtler differences of mood and meaning. In terms of the Poisson superposition above, we should have $N(t) = N_f(t) + N_c(t)$ where these component processes are Markovian but their sum is not. The former process, corresponding to the words of the smaller "functional" subvocabulary, is relatively quickly exhausted, corresponding to having a much larger value rate λ_f: for $t > 5/\lambda_f$, $N_f(t)$ is within about 1% of its limit M_f, and so only the second process is then operative. For all $t \geq 0$ we have from (5.6) that

$$EN_f(t) = M_f(1 - e^{-\lambda_f t}) \quad EN_c(t) = M_c(1 - e^{-\lambda_c t})$$

$$EX(t) = (M_f\lambda_f + M_c\lambda_c)t$$

$$E\Lambda(t) = M_f\lambda_f e^{-\lambda_f t} + M_c\lambda_c e^{-\lambda_c t}$$

$$\approx \lambda_c[M_f + M_c - EN_c(t) - M_f] \approx \lambda_c M(1 - EN(t)/M) \quad (5.16)$$

for $t \geq 5/\lambda_f$. On rewriting this via (5.13) we thus have

$$\alpha_v = \gamma\left(1 - \frac{v}{M}\right) \qquad (5.17)$$

with $\gamma = [M_c\lambda_c/(M_c\lambda_c + M_f\lambda_f)] \times [1 + M_f/M_c]$ effectively equal to the *relative frequency of nonfunctional words*. If this crude categorization of two types of words is expanded to a much larger classification, that is, to a much larger differentiation of frequencies $\{\lambda_j\}$ amongst the nonfunctional words, some more general function than is linear in v may emerge, such as either of the parametrizations at (2.3) and (2.4). The parameter β can be interpreted as a standardized rate of an author's searching for words having the precise shade of meaning wanted in a particular setting, that is, an author's *vocabulary search rate measure*; we may regard its reciprocal as a measure of what is often referred to as *richness*.

Returning to more general distributions for $\{\lambda_j\}$, recall from (5.5) that the distribution of the random variables $T_\nu \equiv X(t'_{\nu+1}) - X(t'_\nu)$ is approximately geometric with parameter α_ν that indicates in an average sense the rate of occurrence of new words when ν have been used to date, the rate being independent of $X(\cdot)$. This means that the total increment Δ_j comes from a first passage property based on a mixture of geometric random variables with slowly changing parameters. Then Δ_j will come from a mixed binomial distribution which, unless it is exceedingly diverse (coming from diverse $\{\lambda_j\}$ and depending also on $N(t)$), will be approximately binomially distributed as indicated in Section 2. The effect of the mixing in general is first to increase and then to decrease the mean at (2.5), and to change the variance term shown there generally in a similar manner.

In general, without more information on the relative sizes of $\{\lambda_j\}$, we are hampered in treating the model in any exact sense. It is still worth observing though that, if in a piece of text an author has a utilized vocabulary V_n that keeps growing with increasing n without evidence of attaining a horizontal asymptote, it is evidence in modeling terms of a long-tailed distribution of rates $\{\lambda_j\}$.

One other family of statistics that has been used concerns counts of words of different frequencies, as for example data based on concordances as in Table 1 of Gani and Sanders (1976). For these data we should consider the model based quantities yielding the expected number of words of frequency r in a text of (time) length t, namely

$$\sum_{j=1}^{M} (\lambda_j)^r \exp(-\lambda_j t)/r! \tag{5.18}$$

The reason for moving from the functional form at (2.3) to (2.4) is embedded in the double inequality, valid for $0 < x < n$,

$$\left(1 - \frac{x}{n}\right)^n < e^{-x} < \left(1 - \frac{x}{n}\right)^{n-x} \tag{5.19}$$

Hence, for M of even moderate size, unless ν is rather closer to M than 0 (see the earlier comments on long-tailed distributions), we deduce from

$$\left(1 - \frac{\nu}{M}\right)^{M\beta} < e^{-\nu\beta} < \left(1 - \frac{\nu}{M}\right)^{M\beta} \bigg/ \left(1 - \frac{\nu}{M}\right)^{\nu\beta} \tag{5.20}$$

that in practical terms, whenever $(1 - \nu/M)^{\nu\beta}$ is close to 1, which holds asymptotically in M for fixed v and β, the parameter M is unidentifiable from the model.

ACKNOWLEDGMENTS

Much of the first and third authors' work was done during a visit to the University of California, Santa Barbara. The second author's research was supported by ONR Contract N00014-86-K-0019.

REFERENCES

Allen, R. F. (1984). *A Stylo-statistical Study of 'Adolphe'*. Slatkine-Champion, Genève et Paris.

Bernet, C. (1983). *Le Vocabulaire des Tragédies de Jean Racine*. Slatkine-Champion, Genève et Paris.

Brainerd, B. (1972). On the relation between types and tokens in literary text. *J. Appl. Probab.*, *9*, 507–518.

Brunet, E. (1978). *Le Vocabulaire de Jean Giraudoux: Structure et Évolution*. Slatkine-Champion, Genève et Paris.

Brunet, E. (1983). *Le Vocabulaire de Proust*. Slatkine-Champion, Genève et Paris.

Brunet, E. (1985). *Le Vocabulaire de Zola*. Slatkine-Champion, Genève et Paris.

Dugast, D. (1979) *Vocabulaire et Stylistique, 1. Théâtre et Dialogue*. Slatkine-Champion, Genève et Paris.

Gani, J. (1975). Stochastic models for type counts in a literary text. In J. Gani (ed.) *Perspectives in Probability and Statistics*, Applied Probability Trust, Sheffield, 313–323.

Gani, J. (1985). Literature and statistics. In N. L. Johnson and S. Kotz (eds.) *Encyclopedia of Statistical Sciences*, Wiley, New York, *5*, 90–95.

Gani, J. and Saunders, I. (1976). Some vocabulary studies of literary texts, *Sankhya, Ser. B.*, *38*, 101–111.

Guiraud, P. (1959). *Problèmes et Méthodes de la Statistique Linguistique*. D. Reidel, Dordrecht,Holland.

Herdan, G. (1960). *Type-token Mathematics: A Textbook of Mathematical Linguistics*. Mouton, The Hague.

Herdan, G. (1964). *Quantitative Linguistics*. Butterworth, London.

Ogden, C. K. (1968). *Basic English: International Second Language*. Harcourt, Brace and World Inc., New York.

Ratkowsky, D. A. (1988). The Travaux de Linguistique Quantitative. *Comp. Humanities 22*, 77–85.

Sichel, H. A. (1986). Word frequency distributions and type-token characteristics. *Math. Scientist, 11*, 45–72.

The Bible in Basic English. (1982). Cambridge University Press, Cambridge.

Yule, G. U. (1944). *The Statistical Study of Literary Vocabulary*. Cambridge University Press, London.

9

A State-Space Approach to Transfer-Function Modeling

P. J. Brockwell and R. A. Davis
Colorado State University, Fort Collins, Colorado

H. Salehi
Michigan State University, East Lansing, Michigan

A state-space realization of the transfer-function model of Box and Jenkins (1976) is used to compute the exact Gaussian likelihood of the input-output series, $\{(X_t, Y_t), t = 1, \ldots, n\}$, and to compute the exact linear mean-square predictor of the output Y_{t+h} based on $\{(X_t, Y_t), t = 1, \ldots, n\}$. The state-space formulation allows model selection using the AIC criterion and the analysis of data with missing values in either or both of the input and output series. An extension of the argument is used to analyze a generalized transfer-function model in which dependence is allowed between the input and output noise sequences. The results are illustrated with reference to the Leading Indicator Sales Data of Box and Jenkins.

1 INTRODUCTION

The term transfer-function modeling, as employed by Box and Jenkins (1976) (see also Priestley (1981) and Brockwell and Davis (1987)), refers to the selection of a model of the form,

$$X_t = \phi_1^{-1}(B)\,\theta_1(B)Z_t \tag{1.1}$$

233

$$Y_t = \phi^{-1}(B)\theta(B)X_t + N_t \tag{1.2}$$

$$N_t = \phi_2^{-1}(B)\phi_2(B)W_t \tag{1.3}$$

to represent a bivariate stationary series $\{(X_t, Y_t)', t = 0, \pm 1, \pm 2, \ldots\}$ of *inputs* X_t and *outputs* Y_t. The inference is based on observations of the two series at times $t = 1, \ldots, n$. In the model (1.1)–(1.3), B denotes the backward shift operator, $\{Z_t\}$ and $\{W_t\}$ are uncorrelated white-noise sequences with means zero and $E(Z_t^2) = \sigma_1^2$, $E(W_t^2) = \sigma_2^2$, and ϕ, ϕ_1, ϕ_2, θ, θ_1, and θ_2 are polynomials of degrees p, p_1, p_2, q, q_1, q_2 respectively. It is assumed that $\phi(z)$, $\phi_1(z)$, and $\phi_2(z)$ are nonzero for all $z \in \mathbb{C}$ such that $|z| \leq 1$ and (without loss of generality) that $\phi(0) = \phi_1(0) = \phi_2(0) = \theta_1(0) = \theta_2(0) = 1$. Notice however that no assumption is made about $\theta(0)$ and that if $\{Y_t\}$ lags behind $\{X_t\}$ by at least one time unit then $\theta(0)$ will be zero.

The usual approach to transfer-function modeling first fits an ARMA model of the form (1.1) to the input series $\{X_t\}$ using a criterion such as the AIC statistic for order selection. This gives estimates ϕ_1 and θ_1 for the polynomials in equation (1.1).

Application of the operator $\theta_1^{-1}(B)\phi_1(B)$ to each side of (1.2) gives

$$Y_t^* = \phi^{-1}(B)\theta(B)Z_t + N_t^* \tag{1.4}$$

where $\{Y_t^*\}$, $\{N_t^*\}$ are the transformed output and noise sequences respectively and $\{Z_t\}$ is uncorrelated with $\{N_t^*\}$.

It is clear from (1.4) that the coefficient t_j of B^j in the power series expansion of $\phi^{-1}(B)\theta(B)$ is

$$t_j = \frac{\text{cov}(Y_t^*, Z_{t-j})}{\text{var}(Z_t)} \tag{1.5}$$

These coefficients can therefore be estimated by applying the *estimated* operator, $\theta_1^{-1}(B)\phi_1(B)$ to the input and output sequences to get estimates of $\{Z_t\}$ and $\{Y_t^*\}$ respectively and then to use these estimated sequences to compute estimates $\hat{t}_j$ from (1.5).

The dependence of the estimated coefficients $\hat{t}_j$ on j is then used to suggest possible polynomials ϕ and θ such that the coefficients of z^j in the power series expansion of $\phi^{-1}(z)\theta(z)$ are approximately $\hat{t}_j$, $j = 0, 1, \ldots$. This provides us with preliminary estimates of the polynomials appearing in (1.2).

More efficient estimation of the parameters appearing in (1.2) and (1.3) is then carried out using least squares as described for example in Brockwell and Davis (1987), Section 12.2. Prediction with the fitted model is carried out using large-sample approximations.

The procedure described above is satisfactory for long series with no missing values. However, it makes use of a number of approximations that for short series can lead to substantial errors in both model-fitting and prediction. The state-space formulation of the model (1.1)–(1.3), which we develop in Section 2, enables us to give an exact treatment of the estimation problem using the Gaussian likelihood of $\{(X_t, Y_t), t = 1, \ldots, n\}$. It also allows us to compute (recursively) the exact linear predictor of Y_{n+h} based on $\{(X_t, Y_t), t = 1, \ldots, n\}$ instead of using a large-sample approximation. In Section 3 we show how to compute the Gaussian likelihood with or without missing observations. Model fitting and order selection can then be carried out using maximum likelihood and the AIC statistic. The calculations are illustrated using the Leading Indicator Sales Data of Box and Jenkins (1976). We also indicate, using this example, how to find the asymptotic distribution of the maximum likelihood estimators of the parameters. In Section 4 we consider the case when differencing is required to make the input and output series stationary.

2 THE STATE SPACE MODEL

Our purpose in this section is to rewrite the model defined by equations (1.1)–(1.3) in "state-space" form. Defining

$$\boldsymbol{\eta}_t := \begin{bmatrix} X_t \\ Y_t \end{bmatrix} \quad \text{and} \quad \boldsymbol{\epsilon}_t := \begin{bmatrix} Z_t \\ W_t \end{bmatrix} \tag{2.1}$$

we shall find matrices A, B, C, and D such that

$$\boldsymbol{\eta}_t = C\boldsymbol{\xi}_t + D\boldsymbol{\epsilon}_t \tag{2.2}$$

where $\{\boldsymbol{\xi}_t\}$ is the unique stationary solution of the "state equation"

$$\boldsymbol{\xi}_{t+1} = A\boldsymbol{\xi}_t + B\boldsymbol{\epsilon}_t \tag{2.3}$$

Once we have found a representation of the form (2.2) and (2.3) it will be a straightforward matter to use the Kalman recursions to compute the best mean square linear predictors $\hat{\eta}_t$ of η_t and their error covariance matrices, $\Delta_t := E[(\eta_t - \hat{\eta}_t)(\eta_t - \hat{\eta}_t)']$.

It is well-known (see for example Aoki (1987)) that the ARMA process $\{X_t\}$ defined by (1.1) has the state-space representation,

$$X_t = C_x \mathbf{x}_t + D_x Z_t \tag{2.4}$$

where $\{\mathbf{x}_t\}$ is the unique stationary solution of

$$\mathbf{x}_{t+1} = A_x \mathbf{x}_t + B_x Z_t \tag{2.5}$$

and

$$A_x = \begin{bmatrix} 0 & 1 & 0 & \cdots & 0 \\ 0 & 0 & 1 & \cdots & 0 \\ \vdots & \vdots & \cdots & \cdots & \vdots \\ 0 & 0 & 0 & \cdots & 1 \\ -\phi_{r_1} & -\phi_{r_1-1} & -\phi_{r_1-2} & \cdots & -\phi_1 \end{bmatrix} \quad B_x = \begin{bmatrix} b_1 \\ b_2 \\ \vdots \\ b_{r_1-1} \\ b_{r_1} \end{bmatrix}$$

$$C_x = [1 \quad 0 \quad \cdots \quad 0] \quad \text{and} \quad D_x = b_0$$

Here $r_1 = \max(p_1, q_1)$ and $\phi_j, b_j, j = 0, \ldots, r_1$, are the coefficients of $z^0, \ldots, z^{r_1}$ in the power series expansions of $\phi_1(z)$ and $\theta_1(z)/\phi_1(z)$ respectively.

It is clear that the noise process $\{N_t\}$ defined by (1.3) has a completely analogous state-space representation which we shall write as

$$N_t = C_n \mathbf{n}_t + D_n W_t \tag{2.6}$$

where $\{\mathbf{n}_t\}$ is the unique stationary solution of

$$\mathbf{n}_{t+1} = A_n \mathbf{n}_t + B_n W_t \tag{2.7}$$

and the matrices A_n, B_n, C_n, and D_n are defined in terms of $\phi_2(\cdot)$ and $\theta_2(\cdot)$ exactly as the matrices A_x, B_x, C_x, and D_x were defined in terms of $\phi_1(\cdot)$ and $\theta_1(\cdot)$.

For the process $\{Y_t\}$ (see (1.2)) we define matrices A_y, B_y, C_y, and D_y in terms of $\phi(\cdot)$ and $\theta(\cdot)$ exactly as A_x, B_x, C_x, and D_x were defined in terms of $\phi_1(\cdot)$ and $\theta_1(\cdot)$. From (1.2) it is then easy to see that $\{Y_t\}$ has the representation

$$Y_t = C_y \mathbf{y}_t + D_y X_t + N_t \tag{2.8}$$

where $\{\mathbf{y}_t\}$ is the unique stationary solution of

$$y_{t+1} = A_y y_t + B_y X_t \tag{2.9}$$

Substituting from (2.4) and (2.6) in the last two equations gives

$$Y_t = C_y y_t + D_y C_x x_t + C_n n_t + D_y D_x Z_t + D_n W_t \tag{2.10}$$

and

$$y_{t+1} = A_y y_t + B_y C_x x_t + B_y D_x Z_t \tag{2.11}$$

The required state-space representation for the bivariate process $\eta_t :=$ $\{(X_t, Y_t)'\}$ can now be written down, by combining the equations (2.4)–(2.7), (2.10) and (2.11), as

$$\begin{bmatrix} X_t \\ Y_t \end{bmatrix} = \begin{bmatrix} C_x & 0 & 0 \\ D_y C_x & C_y & C_n \end{bmatrix} \begin{bmatrix} x_t \\ y_t \\ n_t \end{bmatrix} + \begin{bmatrix} D_x & 0 \\ D_y D_x & D_n \end{bmatrix} \begin{bmatrix} Z_t \\ W_t \end{bmatrix} \tag{2.12}$$

where $\xi_t := \{(x_t', y_t', n_t')'\}$ is the unique stationary solution of

$$\begin{bmatrix} x_{t+1} \\ y_{t+1} \\ n_{t+1} \end{bmatrix} = \begin{bmatrix} A_x & 0 & 0 \\ B_y C_x & A_y & 0 \\ 0 & 0 & A_n \end{bmatrix} \begin{bmatrix} x_t \\ y_t \\ n_t \end{bmatrix} + \begin{bmatrix} B_x & 0 \\ B_y D_x & 0 \\ 0 & B_n \end{bmatrix} \begin{bmatrix} Z_t \\ W_t \end{bmatrix} \tag{2.13}$$

The matrices A, B, C, and D in the representation (2.2) and (2.3) are identifiable by inspection of (2.12) and (2.13). The covariance matrix Σ of $\epsilon_t := [Z_t, W_t]'$ is given by

$$\Sigma = \begin{bmatrix} \sigma_1^2 & 0 \\ 0 & \sigma_2^2 \end{bmatrix} \tag{2.14}$$

and from (2.3) it is easy to compute the covariance matrix Π of the state vector ξ_t as

$$\Pi = \sum_{j=0}^{\infty} A^j B \Sigma B' (A')^j \tag{2.15}$$

We also find from (2.2) that the covariance matrix Γ of $\eta_t := [X_t, Y_t]'$ is

$$\Gamma = C \Pi C' + D \Sigma D' \tag{2.16}$$

Example 1. In Example 12.2.1 of Brockwell and Davis (1987), the Leading Indicator Sales Data of Box and Jenkins (1976) was differenced at lag 1 and mean-corrected. The resulting series $\{(X_t, Y_t)', \ t = 1, \ldots, 149\}$ was then fitted by a model of the form (1.1)–(1.3), namely,

$$X_t = (1 - 0.474B)Z_t, \qquad \{Z_t\} \sim WN(0, 0.0779) \qquad (2.17)$$

$$Y_t = 4.717\,B^3(1 - 0.724B)^{-1}X_t + N_t \qquad\qquad (2.18)$$

$$N_t = (1 - 0.582B)W_t, \qquad \{W_t\} \sim WN(0, 0.0486) \qquad (2.19)$$

Applying the steps described above to reformulate this as a state-space model, we obtain a five-dimensional state vector $\boldsymbol{\xi}_t = [\mathbf{x}_t', \mathbf{y}_t', \mathbf{n}_t']'$ in which $\mathbf{x}_t$ and $\mathbf{n}_t$ are each one-dimensional. Specifically we find that

$$\begin{bmatrix} X_t \\ Y_t \end{bmatrix} = \begin{bmatrix} 1 & 0 & 0 & 0 & 0 \\ 0 & 1 & 0 & 0 & 1 \end{bmatrix} \begin{bmatrix} \mathbf{x}_t \\ \mathbf{y}_t \\ \mathbf{n}_t \end{bmatrix} + \begin{bmatrix} 1 & 0 \\ 0 & 1 \end{bmatrix} \begin{bmatrix} Z_t \\ W_t \end{bmatrix} \qquad (2.20)$$

where $\{\boldsymbol{\eta}_t = (\mathbf{x}_t', \mathbf{y}_t', \mathbf{n}_t')'\}$ is the unique stationary solution of

$$\begin{bmatrix} \mathbf{x}_{t+1} \\ \mathbf{y}_{t+1} \\ \mathbf{n}_{t+1} \end{bmatrix} = \begin{bmatrix} 0 & 0 & 0 & 0 & 0 \\ 0 & 0 & 1 & 0 & 0 \\ 0 & 0 & 0 & 1 & 0 \\ 4.717 & 0 & 0 & 0.724 & 0 \\ 0 & 0 & 0 & 0 & 0 \end{bmatrix} \begin{bmatrix} \mathbf{x}_t \\ \mathbf{y}_t \\ \mathbf{n}_t \end{bmatrix}$$

$$+ \begin{bmatrix} -0.474 & 0 \\ 0 & 0 \\ 0 & 0 \\ 4.717 & 0 \\ 0 & -0.582 \end{bmatrix} \begin{bmatrix} Z_t \\ W_t \end{bmatrix} \qquad (2.21)$$

and $\{Z_t\}$, $\{W_t\}$ are uncorrelated white noise sequences as in (2.17) and (2.19) with variances $\sigma_1^2 = 0.0779$ and $\sigma_2^2 = 0.0486$ respectively.

In Section 3 we shall discuss the question of parameter estimation and model selection and reanalyze this example using maximum Gaussian likelihood and the AIC criterion. Before doing so however we show how the formulation of the model in state-space form facilitates prediction.

2.1 Prediction Using the State-Space Model

A primary purpose of transfer-function modeling is to compute the best linear mean-square predictors $P_t Y_{t+h}$, $h = 1, 2, \ldots$, where P_t denotes projection on $span\{X_s, Y_s, 1 \leq s \leq t\}$. In the notation of (2.2) and (2.3), it suffices to compute $P_t \boldsymbol{\eta}_{t+h}$, $h = 1, 2, \ldots$, where

$$P_t \boldsymbol{\eta}_{t+h} := \begin{bmatrix} P_t X_{t+h} \\ P_t Y_{t+h} \end{bmatrix}, \qquad h = 1, 2, \ldots$$

The Kalman recursions for the state-space model (2.2), (2.3) (see for example Aoki (1987)) enable us to write

$$P_t \boldsymbol{\xi}_{t+1} = A P_{t-1} \boldsymbol{\xi}_t + K_t (\boldsymbol{\eta}_t - P_{t-1} \boldsymbol{\eta}_t) \tag{2.22}$$

$$P_t \boldsymbol{\xi}_{t+h} = A^{h-1} P_t \boldsymbol{\xi}_{t+1}, \qquad h = 2, 3, \ldots \tag{2.23}$$

$$P_t \boldsymbol{\eta}_{t+h} = C P_t \boldsymbol{\xi}_{t+h}, \qquad h = 1, 2, \ldots \tag{2.24}$$

where the Kalman gain, K_t, and the error covariance matrices, $\Delta_t :=$ $E[(\boldsymbol{\eta}_t - P_{t-1} \boldsymbol{\eta}_t)(\boldsymbol{\eta}_t - P_{t-1} \boldsymbol{\eta}_t)']$ and $\Omega_t := E[(\boldsymbol{\xi}_t - P_{t-1} \boldsymbol{\xi}_t)(\boldsymbol{\xi}_t - P_{t-1} \boldsymbol{\xi}_t)']$ are found recursively from the equations,

$$\Omega_1 = \Pi \tag{2.25}$$

$$\Delta_t = C \Omega_t C' + D \Sigma D' \tag{2.26}$$

$$K_t = (A \Omega_t C' + B \Sigma D') \Delta_t^{-1} \tag{2.27}$$

$$\Omega_{t+1} = \Pi - A(\Pi - \Omega_t) A' - K_t \Delta_t K_t' \tag{2.28}$$

and the covariance matrices Σ and Π were defined in (2.14) and (2.15). The one-step predictors $\hat{\boldsymbol{\xi}}_{t+1} := P_t \boldsymbol{\xi}_{t+1}$ and $\hat{\boldsymbol{\eta}}_{t+1} := P_t \boldsymbol{\eta}_{t+1}$, $t = 0$, $1, \ldots$, are found from the recursions,

$$\hat{\boldsymbol{\xi}}_{t+1} = A \hat{\boldsymbol{\xi}}_t + K_t (\boldsymbol{\eta}_t - \hat{\boldsymbol{\eta}}_t) \tag{2.29}$$

$$\hat{\boldsymbol{\eta}}_{t+1} = C \hat{\boldsymbol{\xi}}_{t+1} \tag{2.30}$$

with initial conditions $\boldsymbol{\xi}_1 = \hat{\boldsymbol{\eta}}_1 = 0$. The h-step predictors, $P_t X_{t+h}$ and $P_t Y_{t+h}$, $h = 1, 2, \ldots$, are easily found from the one-step predictors using (2.23) and (2.24). From the relation,

$$\boldsymbol{\xi}_{t+h} - P_t \boldsymbol{\xi}_{t+h} = A(\boldsymbol{\xi}_{t+h-1} - P_t \boldsymbol{\xi}_{t+h-1}) + B \boldsymbol{\epsilon}_{t+h-1}, \qquad h = 2, 3, \ldots$$

we find that $\Omega_t^{(h)} := E[(\boldsymbol{\xi}_{t+h} - P_t \boldsymbol{\xi}_{t+h})(\boldsymbol{\xi}_{t+h} - P_t \boldsymbol{\xi}_{t+h})']$ satisfies the recursions,

$$\Omega_t^{(h)} = A \Omega_t^{(h-1)} A' + B \Sigma B', \qquad h = 2, 3, \ldots \tag{2.31}$$

with $\Omega_t^{(1)} = \Omega_{t+1}$. From (2.2) it then follows that $\Delta_t^{(h)} :=$ $E[(\eta_{t+h} - P_t\eta_{t+h})(\eta_{t+h} - P_t\eta_{t+h})']$ is given by

$$\Delta_t^{(h)} = C\Omega_t^{(h)}C' + D\Sigma D', \qquad h = 1, 2, \ldots \qquad (2.32)$$

Example 1 (continued) For the model introduced in Example 1 we find from (2.14) and (2.15) that

$$\Sigma = \begin{bmatrix} 0.0779 & 0 \\ 0 & 0.0486 \end{bmatrix}$$

and

$$\Pi = \begin{bmatrix} 0.0175 & 0 & 0 & -0.174 & 0 \\ 0 & 1.961 & 0.598 & 0.433 & 0 \\ 0 & 0.598 & 1.961 & 0.598 & 0 \\ -0.174 & 0.433 & 0.598 & 1.961 & 0 \\ 0 & 0 & 0 & 0 & 0.0165 \end{bmatrix}$$

Application of the recursions (2.25)–(2.28) gives the error covariance matrix for the one-step predictor of $[X_{150}, Y_{150}]'$ as

$$\Delta_{150} = \begin{bmatrix} 0.0779 & 0 \\ 0 & 0.0486 \end{bmatrix}$$

The matrix Ω_{150} is found to be zero (to at least seven decimal places). This result is not surprising in view of the fact that ξ_{150} is expressible as a linear combination (with matrix-valued coefficients) of η_s, $s \le 149$. The one-step predictor of $[X_{150}, Y_{150}]'$ is

$$P_{149}\eta_{150} = \begin{bmatrix} 0.138 \\ -0.226 \end{bmatrix}$$

the second component being in close agreement with the results, -0.228, obtained by Brockwell and Davis (1987), p. 461, using approximate large-sample arguments. The error covariance matrix for the two step predictor of $[X_{151}, Y_{151}]'$ is found from (2.31) and (2.32) to be

$$\Delta_{149}^{(2)} = \begin{bmatrix} 0.0954 & 0 \\ 0 & 0.0651 \end{bmatrix}$$

and the predictor itself is found from (2.23) to be

$$P_{149}\eta_{151} = \begin{bmatrix} 0 \\ 0.926 \end{bmatrix}$$

3 MAXIMUM LIKELIHOOD ESTIMATION

Consider the model (1.1)–(1.3) with parameters σ; ν; $\theta_{11}, \ldots, \theta_{1q_1}$; $\phi_{11}, \ldots, \phi_{1p_1}$; $\xi_1, \ldots, \xi_r$; $\phi_1, \ldots, \phi_p$; $\theta_{21}, \ldots, \theta_{2q_2}$; $\phi_{21}, \ldots, \phi_{2p_2}$, where

$$\theta_i(z) = 1 + \theta_{i1} z + \cdots + \theta_{iq_i} z^{q_i}, \qquad i = 1, 2$$

$$\phi_i(z) = 1 + \phi_{i1} z + \cdots + \phi_{ip_i} z_{p_i}, \qquad i = 1, 2$$

$$\theta(z) = z^{b-1}(\xi_1 z + \cdots + \xi_r z^r) = z^{b-1} \xi(z), \qquad b \geq 0, \ r \geq 1$$

and

$$\phi(z) = 1 + \phi_1 z + \cdots + \phi_p z^p$$

To estimate the parameters of the model with specified orders r, p_1, p_2, q, q_1, q_2, and b (the *delay* parameter), we shall maximize the Gaussian likelihood of the data $\{\boldsymbol{\eta}_t = (X_t, Y_t)', t = 1, \ldots, n\}$. The orders will be chosen so as to minimize the AIC statistic.

In order to calculate the exact Gaussian likelihood function based on the observations $\{\boldsymbol{\eta}_1, \ldots, \boldsymbol{\eta}_n\}$, we express it in terms of the innovations as in Schweppe (1965) to obtain

$$L(\boldsymbol{\beta}) = (2\pi)^{-n} \left(\prod_{j=1}^{n} \det \Delta_j \right)^{-1/2} \exp\left[-\frac{1}{2} \sum_{j=1}^{n} (\boldsymbol{\eta}_j - \hat{\boldsymbol{\eta}}_j)' \Delta_j^{-1} (\boldsymbol{\eta}_j - \hat{\boldsymbol{\eta}}_j) \right]$$

(3.1)

where $\boldsymbol{\beta}$ denotes the vector of parameters consisting of all the coefficients and white-noise variances appearing in the model (1.1)–(1.3).

The Gaussian likelihood $L(\boldsymbol{\beta})$ is easily computed for any prescribed set of parameter values from (3.1) and the recursions (2.25)–(2.30). Maximization with respect to the parameters is carried out numerically using a nonlinear optimization algorithm to give the maximum likelihood estimator $\boldsymbol{\beta}$. (For large n a very close approximation to the maximum likelihood estimator $\boldsymbol{\beta}$ can be found by maximizing the marginal likelihood of $X_1, \ldots, X_n$ with respect to $\boldsymbol{\beta}_1$, where $\boldsymbol{\beta}_1$ denotes the parameters in (1.1), and then, with this fixed value for $\boldsymbol{\beta}_1$, maximizing (3.1) with respect to the remaining parameters $\boldsymbol{\beta}_0$ and $\boldsymbol{\beta}_2$ appearing in (1.2) and (1.3) respectively. This procedure amounts to fitting a model to the input series and then maximizing the conditional likelihood of the output series given the input series. It is not completely equivalent to maximizing (3.1) since there is some dependence on $\boldsymbol{\beta}_1$ of the conditional likelihood of $Y_1, \ldots, Y_n$ given $X_1, \ldots, X_n$.)

The likelihood is maximized for a variety of values of p, p_1, p_2, r,

q_1, q_2, b, and the model selected is the one which minimizes the AIC statistic,

$$\text{AIC} = -2 \ln L(\hat{\boldsymbol{\beta}}) + 2(1 + p_1 + p_2 + r + q_1 + q_2) \qquad (3.2)$$

where $\hat{\boldsymbol{\beta}}$ is the maximum likelihood estimator of $\boldsymbol{\beta}$.

Example 1 (continued) Using a model with the same values of r, q, p_1, q_1, p_2, q_2 and b as in (2.17)–(2.19) for the Leading Indicator Sales Data, we find that the maximum Gaussian likelihood model is

$$X_t = (1 - 0.476\,B)\,Z_t, \qquad \{Z_t\} \sim WN(0, 0.0768)$$

$$Y_t = 4.701\,B^3(1 - 0.726\,B)^{-1}X_t + N_t$$

$$N_t = (1 - 0.621B)\,W_t, \qquad \{W_t\} \sim WN(0, 0.0457)$$

Evaluating the right side of (3.2) for this model gives an AIC value of 24.82 (as compared with 25.60 for the model defined by (2.17)–(2.19)). The average of the one-step squared errors $(Y_t - \hat{Y}_t)^2$ is 0.0535 for the model (2.17)–(2.19) and 0.0533 for the maximum likelihood model. There is not a great deal of difference between the two models in this case.

We illustrate the determination of the asymptotic distribution of the maximum likelihood estimators of the coefficients by considering the model fitted to the data in Example 1, namely

$$X_t = (1 - \theta B)\,Z_t$$

$$Y_t = \alpha B^3(1 - \phi B)^{-1}X_t + (1 - \psi B)\,W_t$$

where $|\theta| < 1$, $|\phi| < 1$, $|\psi| < 1$ and $\{Z_t\}$ and $\{W_t\}$ are independent i.i.d. sequences with means zero and variances σ^2 and ν^2 respectively. For the prediction errors $(\boldsymbol{\eta}_j - \hat{\boldsymbol{\eta}}_j)$ and their covariance matrices Δ_j, we find that as $j \to \infty$,

$$\Delta_j \to \begin{bmatrix} \sigma^2 & 0 \\ 0 & \nu^2 \end{bmatrix}$$

and

$$(\boldsymbol{\eta}_j - \hat{\boldsymbol{\eta}}_j) - \begin{bmatrix} Z_j \\ W_j \end{bmatrix} \to 0$$

with the latter holding in both the almost sure and mean square senses. To see the almost sure convergence, first note that

$$\Delta_j \to \begin{bmatrix} \sigma^2 & 0 \\ 0 & \nu^2 \end{bmatrix}$$

at a geometric rate so that

$$\sum_{j=1}^{\infty} \|\eta_{j1} - \hat{\eta}_{j1} - Z_j\|_2 < \infty \quad \text{and} \quad \sum_{j=1}^{\infty} \|\eta_{j2} - \hat{\eta}_{j2} - W_j\|_2 < \infty$$

(where $\|\cdot\|_2$ denotes the L^2 norm). This implies that the series

$$\sum_{j=1}^{\infty} \left(\boldsymbol{\eta}_j - \hat{\boldsymbol{\eta}}_j - \begin{bmatrix} Z_j \\ W_j \end{bmatrix} \right)$$

is a.s. absolutely convergent and hence that the jth summand converges to zero a.s. With probability one we therefore have,

$$-\frac{1}{n} \ln L(\beta) - \ln(2\pi\sigma\nu) - \frac{1}{2n} \left[\sum_{j=1}^{n} \left(\frac{Z_j^2}{\sigma^2} + \frac{W_j^2}{\nu^2} \right) \right] \to 0$$

where

$$Z_j = \sum_{n=0}^{\infty} \theta^n X_{j-n}$$

$$W_j = \sum_{n=0}^{\infty} \psi_n (Y_{j-n} - \phi Y_{j-n-1} - \alpha X_{j-n-3})$$

and $\sum_{n=0}^{\infty} \psi_n z^n = (1 - \phi z)^{-1}(1 - \psi z)^{-1}$, $|z| \le 1$. Correspondingly we obtain for the information matrix based on $\{(X_1, Y_1), \dots, (X_n, Y_n)\}$ (see Brockwell and Davis (1987), Section 8.11, for a similar argument),

$$\frac{1}{n} \left[-E \frac{\partial^2 \ln L(\beta)}{\partial \beta_i \partial \beta_j} \right]_{i,j=1}^{6} \to I$$

where

$$I = \frac{1}{\nu^2} \times \begin{bmatrix} \dfrac{\nu^2}{1 - \theta^2} & 0 & 0 & 0 & 0 & 0 \\ 0 & 2\dfrac{\nu^2}{\sigma^2} & 0 & 0 & 0 & 0 \\ 0 & 0 & E[(U_{t+1} - \phi U_t)^2] & \alpha E[(U_{t+1} - \phi U_t)U_t] & 0 & 0 \\ 0 & 0 & \alpha E[(U_{t+1} - \phi U_t)U_t] & \alpha^2 E(U_t^2) & 0 & 0 \\ 0 & 0 & 0 & 0 & \dfrac{\nu^2}{1 - \psi^2} & 0 \\ 0 & 0 & 0 & 0 & 0 & 2 \end{bmatrix}$$

$$(1 - \phi B)^2 (1 - \psi B) U_t = (1 - \theta B) Z_t$$

and $\beta_1 = \theta$, $\beta_2 = \sigma$, $\beta_3 = \alpha$, $\beta_4 = \phi$, $\beta_5 = \psi$, and $\beta_6 = \nu$.

The asymptotic distribution of the estimated coefficients is found from

$$\sqrt{n} \left(\begin{bmatrix} \hat{\theta} \\ \hat{\alpha} \\ \hat{\phi} \\ \hat{\psi} \end{bmatrix} - \begin{bmatrix} \theta \\ \alpha \\ \phi \\ \psi \end{bmatrix} \right) \Rightarrow N(0, \Sigma) \qquad (3.3)$$

where

$$\Sigma^{-1} = \nu^{-2}$$

$$\times \begin{bmatrix} \nu^2 (1 - \theta^2)^{-1} & 0 & 0 & 0 \\ 0 & E[(U_{t+1} - \phi U_t)^2] & \alpha E[(U_{t+1} - \phi U_t)U_t] & 0 \\ 0 & \alpha E[(U_{t+1} - \phi U_t)U_t] & \alpha^2 E(U_t^2) & 0 \\ 0 & 0 & 0 & \nu^2 (1 - \psi^2)^{-1} \end{bmatrix}$$

Estimating Σ by substituting the estimated parameter values from Example 1 gives

$$\hat{\Sigma} = \begin{bmatrix} 0.7753 & 0 & 0 & 0 \\ 0 & 0.3448 & -0.01757 & 0 \\ 0 & -0.01757 & 0.001954 & 0 \\ 0 & 0 & 0 & 0.6144 \end{bmatrix}$$

3.1 Missing Values

Missing values in one or both of the series $\{X_t\}$, $\{Y_t\}$ can be handled using the following device. Define the 2×2 matrices,

$$I = \begin{bmatrix} 1 & 0 \\ 0 & 1 \end{bmatrix}, \quad E_1 = \begin{bmatrix} 1 & 0 \\ 0 & 0 \end{bmatrix}, \quad E_2 = \begin{bmatrix} 0 & 0 \\ 0 & 1 \end{bmatrix}, \quad O = \begin{bmatrix} 0 & 0 \\ 0 & 0 \end{bmatrix}$$

and let

$$M_t := \begin{cases} I & \text{if } \eta_t = (X_t, Y_t)' \text{ is missing} \\ E_1 & \text{if } X_t \text{ is missing and } Y_t \text{ is not} \\ E_2 & \text{if } Y_t \text{ is missing and } X_t \text{ is not} \\ O & \text{otherwise} \end{cases}$$

With $\{\xi_t\}$ defined by (2.3) exactly as before, we now define a new sequence of observation vectors,

$$\eta_t = C_t \xi_t + D_t \epsilon_t + M_t \delta_t \tag{3.4}$$

where $C_t := (I - M_t)C$, $D_t := (I - M_t)D$ and $\{\delta_t\}$ is an i.i.d. sequence of bivariate $N(0, I)$ random vectors independent of $\{\epsilon_t\}$. The sequence $\{\eta_t\}$ is then precisely the sequence defined by the observation equation (2.2), except that unobserved values are replaced in the sequence (3.4) by independent standard normal random variables. To within multiplication by a parameter-independent factor, the likelihood of the sequence defined by (3.4) is therefore the same as the likelihood of the *observed* values of X_t and Y_t. Since the realized numerical values of the vectors δ_t enter the likelihood as multiplicative parameter, independent constants only, we can assume (for likelihood maximization) that they are all zero. Thus we replace all missing values by zeroes in order to generate a complete set of "observations."

The Gaussian likelihood of the original data can thus be computed from (3.1) (to within multiplication by a parameter-independent factor) by replacing all missing values with zeroes and using the observation equation (3.4) in place of (2.2). The appropriate modifications of the Kalman recursions (2.25)–(2.30) required to perform these calculations are as follows:

$$\Omega_1 = \Pi \tag{3.5}$$

$$\Delta_t = C_t \Omega_t C_t' + D_t \Sigma D_t' + M_t M_t' \tag{3.6}$$

$$K_t = (A\Omega_t C_t' + B\Sigma D_t') \Delta_t^{-1} \tag{3.7}$$

$$\Omega_{t+1} = \Pi - A(\Pi - \Omega_t)A' - K_t \Delta_t K_t' \tag{3.8}$$

where the covariance matrices Σ and Π were defined in (2.14) and (2.15). The one-step predictors $\hat{\xi}_{t+1} := P_t \xi_{t+1}$ and $\hat{\eta}_{t+1} := P_t \eta_{t+1}$, $t = 0, 1, \ldots$, are found from the recursions,

$$\hat{\xi}_{t+1} = A\hat{\xi}_t + K_t(\eta_t - \hat{\eta}_t) \tag{3.9}$$

$$\hat{\eta}_{t+1} = C_{t+1}\hat{\xi}_{t+1} \tag{3.10}$$

with initial conditions $\hat{\xi}_1 = \hat{\eta}_1 = 0$.

As in the case when there are no missing values, a very good appoximation to the maximum likelihood estimaors can be obtained by first maximizing the likelihood of the input series with respect to β_1 (dealing with missing values as described in Brockwell and Davis (1987), Section 12.3) and then maximizing the Gaussian likelihood of the bivariate series with respect to β_0 and β_2, keeping β_1 fixed.

Example 1 (continued) Deleting the last three sales figures from the Sales Leading Indicator Data of Box and Jenkins (1976), we can use the technique described above to maximize the Gaussian likelihood of the remaining 295 differenced observations, treating the last three differenced sales figures as missing observations. Using the same values for r, q, p_1, q_1, p_2, q_2 and b as before, we obtain the maximum likelihood model,

$$X_t = (1 - 0.476B)Z_t, \qquad \{Z_t\} \sim WN(0, 0.0768)$$

$$Y_t = 4.707\,B^3(1 - 0.725\,V)^{-1}X_t + N_t$$

$$N_t = (1 - 0.624B)W_t, \qquad \{W_t\} \sim WN(0, 0.0462)$$

This model can now be used to find the best linear least squares estimates (based on all the available observations) of both the missing three differenced sales observations and of future values. The following table shows these estimates as computed from the missing-value model. The last column of the table shows the *actual* values of the differenced sales figures at times $t = 148$, 149, and 150, as well as the predicted differenced sales values for an additional three time units, obtained from the maximum likelihood model for the original data with no missing observations. The consistency between the two sets of predicted values demonstrates that little would be lost (as far as sales prediction is concerned) if the last three sales figures in this example were missing.

4 NONSTATIONARY INPUT AND OUTPUT PROCESSES

Finally we indicate how a state-space model can be constructed for the *original* observations when differencing must be performed before modeling the data as a bivariate stationary series (as in Example 1). Let

Table 9.1 Linear least squares estimates of differenced sales

t	Missing values at $t = 148, 149, 150$	No missing values
148	−1.178	−1.000
149	0.380	0.400
150	0.663	0.500
151	0.162	0.188
152	1.353	1.341
153	−0.750	−0.757

$\{\zeta_t, t = 0, 1, 2, \ldots\}$ denote the original series and let $\eta_t = \zeta_t - \zeta_{t-1} - \mu$, $t = 1, 2, \ldots$, where μ is the sample mean of the differenced data. If the differenced and mean corrected sequence $\{\eta_t\}$ has a representation of the form (2.2) and (2.3) with coefficient matrices A, B, C and D, and if we introduce the enlarged state vectors $\mathbf{S}_t = (\xi_t', \zeta_{t-1}' - (t - 1)\mu')'$, we find that the sequence $\{\zeta_t - t\mu\}$ has the representation

$$\zeta_t - t\mu = [A \quad I]\mathbf{S}_t + B\epsilon_t \tag{4.1}$$

with state equation

$$\mathbf{S}_{t+1} = \begin{bmatrix} A & 0 \\ C & I \end{bmatrix}\mathbf{S}_t + \begin{bmatrix} B \\ D \end{bmatrix}\epsilon_t \tag{4.2}$$

and initial condition, $\mathbf{S}_1 = (\xi_1', \zeta_0')'$ where $\xi_1 = \sum_{j=0}^{\infty} A^j\epsilon_{-j}$. Assuming that ζ_0 is orthogonal to $\{\epsilon_t, t = 0, \pm1, \ldots\}$, the Kalman recursions can now be used to predict the original data with and without missing values and to compute the Gaussian likelihood of the data conditional on ζ_0.

ACKNOWLEDGMENT

This work was supported in part by WSF grants DMS 8802559 and DMS 9006422.

REFERENCES

Aoki, M. (1987). *State Space Modeling of Time Series*. Springer-Verlag, Berlin.

Box, G. E. P., and Jenkins, G. M. (1976). *Time Series Analysis: Forecasting and Control*. 2nd edition. Holden Day, San Francisco.

Brockwell, P. J., and Davis, R. A. (1987). *Time Series: Theory and Methods*. Spinger-Verlag, New York.

Priestley, M. B. (1981). *Spectral Analysis and Time Series*. Vols. 1 and 2. Academic Press, New York.

Schweppe, F. C. (1965). Evaluation of likelihood functions for Gaussian signals, *IEEE Transactions on Information Theory*, IT-11, 61–70.

10
Shrinkage Estimation for a Dynamic Input-Output Linear Model

Young-Won Kim, David M. Nickerson, and I. V. Basawa
University of Georgia, Athens, Georgia

A shrinkage estimator for the regression parameter in a dynamic input-output linear model is presented as an alternative to the maximum likelihood estimator (MLE). It is shown that the shrinkage estimator has a lower risk than the maximum likelihood estimator when the coefficient of the lagged dependent variable is known. Further, when the lagging coefficient is unknown, we show that the asymptotic risk of the shrinkage estimator is less than that of the maximum likelihood estimator at and in the neighborhood of the shrinkage point, whereas the asymptotic risks for the two estimators are the same at all other fixed parameter values.

1 INTRODUCTION

Consider the following dynamic input-output model containing fixed independent variables and also lagged values of the dependent variable specified by

$$Y_t = \phi Y_{t-1} + X_t^T \beta + \epsilon_t \tag{1.1}$$

where $\{\epsilon_t\}$, $t = 1, 2, \ldots, n$, are independent unobservable $N(0, \sigma^2)$ random variables, $\{Y_t\}$ are observable random variables or responses, $\{X_t\}$ are known ($p \times 1$) vectors of covariates or independent variables, β is an unknown ($p \times 1$) vector of regression parameters, and ϕ is the autoregressive parameter. Further, fix $Y_0 = 0$. If we denote $Y = (Y_1, \ldots, Y_n)^T$, $Y_l = (Y_0, Y_1, \ldots, Y_{n-1})^T$, $X = (X_1, \ldots, X_n)^T$ and $\epsilon = (\epsilon_1, \ldots, \epsilon_n)^T$ the above model can be writen as

$$Y = \phi Y_l + X\beta + \epsilon \qquad (1.2)$$

where $\epsilon \sim N(0, \sigma^2 I_n)$ and I_n is the identity matrix of order $n \times n$. The symbol $\sim$ here denotes distribution equivalence.

The properties of the maximum likelihood estimate of β in (1.2) are well-known. Our main goal in this paper is to study a Stein type shrinkage estimate of β which beats the maximum likelihood estimate in small as well as large samples in a sense to be specified later. First, let us consider the maximum likelihood estimation. If ϕ is known, define

$$Z(\phi) = Y - \phi Y_l \qquad (1.3)$$

so that

$$Z(\phi) = X\beta + \epsilon \qquad \text{and} \qquad Z(\phi) \sim N(X\beta, \sigma^2 I_n) \qquad (1.4)$$

Further, if we define the matrix $A(\phi)$ as

$$A(\phi) = \begin{bmatrix} 1 & 0 & 0 & \cdots & 0 & 0 \\ -\phi & 1 & 0 & \cdots & 0 & 0 \\ 0 & -\phi & 1 & \cdots & 0 & 0 \\ \cdot & \cdot & \cdot & & \cdot & \cdot \\ \cdot & \cdot & \cdot & & \cdot & \cdot \\ \cdot & \cdot & \cdot & & \cdot & \cdot \\ 0 & 0 & 0 & \cdots & -\phi & 1 \end{bmatrix} \qquad (1.5)$$

then

$$A^{-1}(\phi) = \begin{bmatrix} 1 & 0 & 0 & \cdots & 0 \\ \phi & 1 & 0 & \cdots & 0 \\ \phi^2 & \phi & 1 & \cdots & 0 \\ \cdot & \cdot & \cdot & & \cdot \\ \cdot & \cdot & \cdot & & \cdot \\ \cdot & \cdot & \cdot & & \cdot \\ \phi^{n-1} & \phi^{n-2} & \phi^{n-3} & \cdots & 1 \end{bmatrix} = B(\phi) \text{ (say)} \qquad (1.6)$$

and (1.3) can be written as

$$Z(\phi) = A(\phi)Y \tag{1.7}$$

Consequently

$$Y = A^{-1}(\phi)Z(\phi) = B(\phi)Z(\phi) \tag{1.8}$$

and for given β, we have

$$Y|\beta \sim N(B(\phi)X\beta, \sigma^2 B(\phi)B^T(\phi)) \tag{1.9}$$

From (1.9), for known ϕ, the maximum likelihood estimator of β is given by

$$\begin{aligned} \hat{\beta}_n(\phi) &= (X^T X)^{-1} X^T B^{-1}(\phi) Y \\ &= (X^T X)^{-1} X^T A(\phi) Y \\ &= (X^T X)^{-1} X^T Z(\phi) \end{aligned} \tag{1.10}$$

It follows that

$$\hat{\beta}_n(\phi) \sim N(\beta, \sigma^2 (X^T X)^{-1}) \tag{1.11}$$

where it is assumed that the rank of $X^T X$ is p.

We now consider a shrinkage estimator $\delta_n^s(\phi)$ of the type

$$\delta_n^s(\phi) = \lambda + (I_p - C_n(\phi))(\hat{\beta}_n(\phi) - \lambda) \tag{1.12}$$

where λ is a known $(p \times 1)$ vector, I_p is the $(p \times p)$ identity matrix, and $C_n(\phi)$ is a certain $(p \times p)$ matrix to be specified later. If ϕ is unknown we need only replace ϕ in (1.11) and (1.13) by its maximum likelihood estimate $\hat{\phi}_n$ where

$$\hat{\phi}_n = \frac{Y_l'(I_n - H)Y}{Y_l'(I_n - H)Y_l} \tag{1.13}$$

with

$$H = X(X^T X)^{-1} X^T \tag{1.14}$$

Denote $\hat{\beta}_n(\hat{\phi}_n)$ and $\delta_n^s(\hat{\phi}_n)$ by $\hat{\beta}_n$ and δ_n^s, respectively. It will be shown that for large samples δ_n^s is more efficient than $\hat{\beta}_n$ in a sense to be defined later.

In Section 2 we derive the shrinkage estimator as an empirical Bayes estimator. For the case $\phi = 0$, Sclove (1968) established the dominance of $\delta_n^s(0)$ over $\hat{\beta}_n(0)$ for any $n \geq p + 1$ and $p \geq 3$. In Section 3, we extend Sclove's result to the case $\phi \neq 0$ and ϕ known. In Section 4, we consider the cases both when ϕ is known and when it is unknown and study the asymptotic properties of $\delta_n^s(\phi)$ and δ_n^s, and compare them with $\hat{\beta}_n(\phi)$ and $\hat{\beta}_n$ respectively.

For early work in the case of independent observations, (i.e., $\phi = 0$), and for special types of the covariance matrix X, see James and Stein (1961) when σ^2 known; Berger (1976) when the covariance matrix of ϵ is known, but not necessarily I_p; and Efron and Morris (1976) when σ^2 is unknown. For a general class of estimators dominating the MLE, see Baranchik (1970), Strawderman (1971), Efron and Morris (1976) and Berger (1976). Further, for the case in which the covariance matrix of ϵ is diagonal, but not necessarily a multiple of I_p see Berger and Bock (1976a). Finally, for the case in which the covariance matrix of ϵ is entirely unknown see Berger and Bock (1976b), Berger et al. (1977) and Berger and Haff (1983).

What needs to be stressed at this point is that the key difference between previous work and the present paper is the lack of independence of the estimators of the target parameters and the nuisance parameters when the autoregressive parameter is unknown. When the autoregressive parameter is zero or known, it is the independence that is exploited in previous work, via Stein's lemma, which allows for finite sample dominance of the shrinkage estimator over the MLE. It happens that the dependence is the stumbling block which makes it difficult for anything other than large sample results for the case in which the autoregressive parameter is unknown.

2 DERIVATION OF THE SHRINKAGE ESTIMATOR AS AN EMPIRICAL BAYES ESTIMATOR

First, suppose that the loss in estimating β by δ_n, a function of Y and X, is represented by

$$L_n(\delta_n, \beta) = (\delta_n - \beta)^T Q_n(\delta_n - \beta) \qquad (2.1)$$

where Q_n is a ($p \times p$) positive definite weight matrix to be chosen by the statistician. If Q_n is chosen to be I_p, then (2.1) reduces to

$$L_n(\delta_n, \beta) = \Sigma_{i=1}^p (\delta_{ni} - \beta_i)^2$$

the usual sum of squares of the componentwise errors. Another useful choice for Q_n is $Q_n = n^{-1}(X^T X)$, in which case then (2.1) reduces to

$$L_n(\delta_n, \beta) = \frac{1}{n}(\delta_n - \beta)^T X^T X (\delta_n - \beta)$$

$$= \frac{1}{n}(B(\phi)X\delta_n - B(\phi)X\beta)^T (B(\phi)B^T(\phi))^{-1}$$

$$\times (B(\phi)X\delta_n - B(\phi)X\beta)$$

$$= \frac{1}{n}(\hat{Y} - E(Y))^T (B(\phi)B^T(\phi))^{-1}(\hat{Y} - E(Y)) \quad (2.3)$$

where $\hat{Y} = B(\phi)X\delta_n$. Recalling the distribution of Y from (1.9), (2.3) reduces to the usual weighted mean squared error in prediction.

As an alternative to the usual maximum likelihood estimator $\hat{\beta}_n(\phi)$ in (1.10) consider the following shrinkage estimator

$$\delta_n^s(\phi) = \lambda + \left\{ I_p - \frac{b\hat{\sigma}_n^2(\phi)}{\|\hat{\beta}_n(\phi) - \lambda\|^2} Q_n^{-1} \Sigma_n \right\}(\hat{\beta}_n(\phi) - \lambda) \quad (2.4)$$

where λ is a predetermined vector, b is a real number, $b \in \{0, 2(p - 2)\}$,

$$\|a\|^2 = a^T \Sigma_n Q_n^{-1} \Sigma_n a \quad (2.5)$$

$\Sigma_n = X^T X$ (for ϕ known and will subsequently be adjusted for ϕ unknown) and $\hat{\sigma}_n^2(\phi)$ is an appropriate estimate of σ^2 to be defined later. Note that for the specific choice $Q_n = n^{-1}(X^T X)$, (2.4) reduces to

$$\delta_n^s(\phi) = \lambda + (1 - \hat{d}_n(\phi))(\hat{\beta}_n(\phi) - \lambda) \quad (2.6)$$

where

$$\hat{d}_n(\phi) = b \frac{\hat{\sigma}_n^2(\phi)}{(\hat{\beta}_n(\phi) - \lambda)^T X^T X (\hat{\beta}_n(\phi) - \lambda)} \quad (2.7)$$

a scalar. In the next section, we show that $\delta_n^s(\phi)$ dominates $\hat{\beta}_n(\phi)$ for every $n \geq p + 1$ in the sense of having a smaller risk function. In this section we shall give an empirical Bayes motivation for the estimator in (2.6).

Suppose β is a random variable with a normal prior having mean λ, assumed known, and covariance matrix $\tau^2(X^T X)^{-1}$. Then the posterior distribution of β given $Y = y$ is normal with mean vector μ and covariance matrix Σ where

$$\mu = \lambda + \left\{ 1 - \frac{\sigma^2}{\tau^2 + \sigma^2} \right\}(\hat{\beta}_n(\phi) - \lambda) \quad \text{and} \quad \Sigma = \sigma^2 \frac{\tau^2}{\tau^2 + \sigma^2}(X^T X)^{-1}$$

$$(2.8)$$

Since the loss function (2.1) is weighted squared error the Bayes estimator of β is given by

$$\delta_n^B(\phi) = \mu = \lambda + \left\{1 - \frac{\sigma^2}{\tau^2 + \sigma^2}\right\}(\hat{\beta}_n(\phi) - \lambda) \qquad (2.9)$$

Further, suppose σ^2 and τ^2 are unknown. Then, using the marginal distribution of Y, we shall develop estimators of σ^2 and $(\sigma^2 + \tau^2)$. Then these estimators are substituted in (2.8). The resulting estimate will be referred to as an empirical Bayes estimate.

Now, marginally, $Z(\phi) = Y - \phi Y_l$ is distributed as normal with mean vector $X\lambda$ and covariance matrix $(\sigma^2 I_n + \tau^2 H)$. Consequently, the marginal distribution of $\hat{\beta}_n(\phi)$ is normal with mean vector λ and covariance matrix $(\tau^2 + \sigma^2)(X^T X)^{-1}$. Hence, marginally,

$$(\hat{\beta}_n(\phi) - \lambda)^T (X^T X)(\hat{\beta}_n(\phi) - \lambda) \sim (\tau^2 + \sigma^2)\chi_p^2.$$

Therefore, an unbiased estimator of $(\tau^2 + \sigma^2)^{-1}$ is given by

$$(p - 2)/\{(\hat{\beta}_n(\phi) - \lambda)^T (X^T X)(\hat{\beta}_n(\phi) - \lambda)\} \qquad (2.10)$$

Next, consider the estimate $\hat{\sigma}_n^2(\phi)$ of σ^2 defined by

$$\begin{aligned}
\hat{\sigma}_n^2(\phi) &= \frac{1}{n - p}(Z_n(\phi) - X\hat{\beta}_n(\phi))^T (Z_n(\phi) - X\hat{\beta}_n(\phi)) \\
&= \frac{Z_n^T(\phi)(I_n - H)^T (I_n - H)Z_n(\phi)}{(n - p)} \\
&= \frac{(Z_n(\phi) - X\lambda)^T (I_n - H)(Z_n(\phi) - X\lambda)}{(n - p)} \qquad (2.11)
\end{aligned}$$

Recalling that $Z_n(\phi) \sim N(X\lambda, \sigma^2 I_n + \tau^2 H)$ and using the facts that

$$\frac{1}{\sigma^2}(I_n - H)(\sigma^2 I_n + \tau^2 H) = (I_n - H)$$

this last matrix being idempotent with rank $r(I_n - H) = n - p$, and $(X\lambda)^T (I_n - H)(X\lambda) = 0$, we have that

$$\hat{\sigma}_n^2(\phi) \sim \frac{\sigma^2}{n - p}\chi_{n-p}^2$$

Hence, $\hat{\sigma}_n^2(\phi)$ serves as an unbiased estimate of σ^2. Also, note that

$$(\hat{\beta}_n(\phi) - \lambda)^T (X^T X)(\hat{\beta}_n(\phi) - \lambda) = (Z(\phi) - X\lambda)^T H(Z(\phi) - X\lambda)$$

and

$$H(\sigma^2 I_n + \tau^2 H)(I_n - H) = 0$$

Consequently, the estimate of $(\tau^2 + \sigma^2)^{-1}$ in (2.10) and the estimate of σ^2 in (2.11) are independently distributed. Therefore, an unbiased estimate of $d = \sigma^2/(\tau^2 + \sigma^2)$ is given by

$$\hat{d}(\phi) = (p - 2) \frac{\hat{\sigma}_n^2(\phi)}{(\hat{\beta}_n(\phi) - \lambda)^T (X^T X)(\hat{\beta}_n(\phi) - \lambda)} \qquad (2.12)$$

Now, substituting (2.12) for $\sigma^2/(\tau^2 + \sigma^2)$ in (2.8) we obtain the empirical Bayes estimate which coincides with $\delta_n^s(\phi)$ in (2.6) with $b = (p - 2)$.

The above derivation gives an empirical Bayes motivation for the shrinkage estimate (2.6). From this point forward we shall return to the frequency framework with the conditional distribution of Y given β, and use this to assess the properties of the shrinkage estimate and compare them with those of the maximum likelihood estimator $\hat{\beta}_n(\phi)$ where ϕ is known and unknown.

3 RISK DOMINANCE OF THE SHRINKAGE ESTIMATOR FOR FINITE SAMPLES WHEN ϕ IS KNOWN

The results in this section are a simple extension of those of Sclove (1968) for $\phi = 0$, to the dependent situation with $\phi \neq 0$ and ϕ known. Recall that $\hat{\beta}_n(\phi)$ and $\delta_n^s(\phi)$ given in (1.10) and (2.4) respectively denote the maximum likelihood and shrinkage estimators. The following theorem states that the risk function of $\delta_n^s(\phi)$ is smaller than that of $\hat{\beta}_n(\phi)$ for every $n \geq p + 1$. The basic reference model in this section is the conditional distribution of Y given β.

At this point it should be noted that the proof of the following theorem is a simple adaptation of the results of Sclove (1968). However, since certain steps in the proof will be used in later sections, we shall present a brief outline of the required arguments.

Theorem 3.1 Let the loss function $L_n(\delta_n, \beta)$ be defined by (2.1), and suppose $R(\delta_n^s(\phi), \beta) = E\{L_n(\delta_n^s(\phi), \beta)\}$ and $R(\hat{\beta}_n(\phi), \beta) = E\{L_n(\hat{\beta}_n(\phi), \beta)\}$ denote the risk functions of the shrinkage and maximum likelihood estimators, respectively. Then for ϕ known and, for all β,

$$R(\delta_n^s(\phi), \beta) < R(\hat{\beta}_n(\phi), \beta) \text{ for } b \in \left\{0, 2(p-2)\frac{n-p}{n-p+2}\right\} \quad (3.1)$$

Proof: Using the definition of $\delta_n^s(\phi)$ in (2.4), we have

$$R(\delta_n^s(\phi), \beta) = R(\hat{\beta}_n(\phi), \beta)$$

$$+ b^2 E\left\{\frac{\hat{\sigma}_n^4(\phi)}{\|\hat{\beta}_n(\phi) - \lambda\|^2}\right\}$$

$$- 2bE\left\{\frac{\hat{\sigma}^2(\phi)}{\|\hat{\beta}_n(\phi) - \lambda\|^2}(\hat{\beta}_n(\phi) - \lambda)^T X^T X (\hat{\beta}_n(\phi) - \beta)\right\}$$

$$(3.2)$$

By the independence of $\hat{\sigma}_n^2(\phi)$ and $\hat{\beta}_n(\phi)$, and by Stein's identity we have

$$R(\delta_n^s(\phi), \beta) = R(\hat{\beta}_n(\phi), \beta)$$

$$- \frac{n-p+2}{n-p}\sigma^4 E\left\{\|\hat{\beta}_n(\phi) - \lambda\|^{-2}\right\}$$

$$\times b\left\{2(p-2)\frac{n-p}{n-p+2} - b\right\}$$

$$< R(\hat{\beta}_n(\phi), \beta) \quad \text{for all } b \in \left\{0, 2(p-2)\frac{n-p}{n-p+2}\right\}$$

$$(3.3)$$

4   ASYMPTOTIC PROPERTIES OF THE SHRINKAGE ESTIMATOR

4.1. Autoregressive Parameter Known

First, consider the case when ϕ is known. Using the notation of Section 3 we shall derive the limiting risks of the maximum likelihood and shrinkage estimators, and then compare the two estimators.

Theorem 4.1 Assume that

$$Q_n \to A \quad (4.1)$$

and

$$n^{-1}X^T X \to B \tag{4.2}$$

as $n \to \infty$, where A and B are both positive definite matrices. Let $\hat{\beta}_n(\phi)$ and $\delta_n^s(\phi)$ be the estimates defined in (1.11) and (2.4), respectively. We then have, as $n \to \infty$,

   (i) $R(\hat{\beta}_n(\phi), \beta) = an^{-1} + o(n^{-1})$

   (ii) $\{R(\hat{\beta}_n(\phi), \beta) - R(\delta_n^s(\phi)m\,\beta)\}$
      $= bn^{-1} + o(n^{-1}), \quad$ for $\beta = \lambda$

   (iii) $\{R(\hat{\beta}_n(\phi), \beta) - R(\delta_n^s(\phi), \beta)\} = cn^{-1} + o(n^{-1})$,
      for $\beta = \lambda + n^{-1/2}\gamma$

where γ is a $(p \times 1)$ vector of real numbers, and

   (iv) $\{R(\hat{\beta}_n(\phi), \beta) - R(\delta_n^s, \beta)\} = dn^{-2} + o(n^{-2})$,
      for fixed $\beta \neq \lambda$

where

$$a = \sigma^2 \operatorname{tr}\{AB^{-1}\} \tag{4.3}$$

$$b = \sigma^2 b\{2(p - 2) - b\} \int_0^\infty |I_p + 2tA^{-1}B|^{-1/2}\, dt \tag{4.4}$$

$$c = \sigma^4 b\{2(p - 2) - b\} E\{W^T BA^{-1}BW\}^{-1} \tag{4.5}$$

with $W \sim N(\gamma, \sigma^2 B^{-1})$, and

$$d = \sigma^4 b\{2(p - 2) - b\}\{(\beta - \lambda)^T BA^{-1}B(\beta - \lambda)\}^{-1} \tag{4.6}$$

Note that in the special case when $Q_n = n^{-1}X^T X$, the matrix A in the above theorem is identical with B, and consequently the above expressions for a, b, c, and d simplify.

Proof: (i) We have

$$nR(\hat{\beta}_n(\phi), \beta) = nE\{(\hat{\beta}_n(\phi) - \beta)^T Q_n(\hat{\beta}_n(\phi) - \beta)\}$$

$$= n\operatorname{tr}[Q_n E\{\hat{\beta}_n(\phi) - \beta)(\hat{\beta}_n(\phi) - \beta)^T\}]$$

$$= \sigma^2 \operatorname{tr}[Q_n n(X^T X)^{-1}]$$

$$\to \sigma^2 \operatorname{tr}\{AB^{-1}\}$$

since $Q_n \to A$, and $n^{-1}(X^T X) \to B$, as $n \to \infty$.

(ii) This follows from (iii) if $\gamma = 0$.

(iii) We have from (3.3)

$$(R(\hat{\beta}_n(\phi), \beta) - R(\delta_n^s(\phi), \beta)\}$$

$$= \frac{n - p + 2}{n - p} \sigma^4 E\{\|\hat{\beta}_n(\phi) - \lambda\|^{-2}\}b\left\{2(p - 2)\left(\frac{n - p}{n - p + 2}\right) - b\right\} (4.7)$$

Next, consider $E\{\|\hat{\beta}_n(\phi) - \lambda\|^{-2}\}$ in (4.7). To evaluate this term let us consider a matrix P_n such that $P_n(X^T X)^{-1} P_n^T = D_n^{-1} =$ diag$(d_{1n}^{-1}, \ldots, d_{pn}^{-1})$ and $P_n Q_n^{-1} P_n^T = I_p$. The existence of a matrix P_n such as this is given by the simultaneous diagonalization theorem (see, Rao (1973), page 41). Then, defining $L_n = P_n(\hat{\beta}(\phi) - \lambda) \sim N(\xi_n, \sigma^2 D_n^{-1})$ with $\xi_n = R_n(\beta - \lambda)$, we have

$$E\{\|\hat{\beta}_n(\phi) - \lambda\|^{-2}\} = E\left\{\left(\sum_{j=1}^{P} L_{jn}^2 d_{jn}^2\right)^{-1}\right\}$$

$$= \sigma^{-2} \int_0^\infty E\left\{\exp\left(-t \sum_{j=1}^{P} \frac{L_{jn}^2}{\sigma^2} d_{jn}^2\right)\right\} dt = \sigma^{-2} \int_0^\infty \prod_{j=1}^{P} E\left\{\exp\left(-\frac{L_{jn}^2}{\sigma^2} d_{jn}^2\right)\right\} dt$$

$$= \sigma^{-2} \int_0^\infty |I_p + 2tD_n|^{-1/2} \times \exp\{-\sigma^{-2} t \xi_n^T D_n (I_p + 2tD_n)^{-1} D_n \xi_n\} dt$$

$$= \sigma^{-2} \int_0^\infty |I_p + 2tQ_n^{-1}(X^T X)|^{-1/2} \times \exp\{-\sigma^{-2} t(\beta - \lambda)^T (X^T X)$$

$$\times (Q_n + 2t(X^T X))^{-1} (X^T X)(\beta - \lambda)\} dt \qquad (4.8)$$

Using the fact that $\beta = \lambda + n^{-1/2}\gamma$ in (4.8) we obtain

$$nE\{\|\hat{\beta}_n(\phi) - \lambda\|^{-2}\}$$

$$= \frac{n}{\sigma^2} \int_0^\infty |I_p + 2tQ_n^{-1}(X^T X)|^{-1/2} \exp\{-\sigma^{-2} tn^{-1} \gamma^T (X^T X)$$

$$\times (Q_n + 2t(X^T X))^{-1} (X^T X) \gamma\} dt$$

$$= \sigma^{-2} \int_0^\infty |I_p + 2tQ_n^{-1} n^{-1} (X^T X)|^{-1/2} \exp\{-\sigma^{-2} t \gamma^T n^{-1}(X^T X)$$

$$\times (Q_n + 2tn^{-1}(X^T X))^{-1} n^{-1}(X^T X)\,\gamma\}\,dt$$

$$\to \sigma^{-2} \int_0^\infty |I_p + 2tA^{-1}B|^{-1/2} \exp\{-\sigma^{-2}t\gamma^T B(A + 2tB)^{-1} B\gamma\}\,dt.$$

$$(4.9)$$

Now, with $W \sim N(\gamma, \sigma^2 B^{-1})$, we have

$$E\{(W^T BA^{-1} BW/\sigma^2)^{-1}\}$$

$$= \int_0^\infty E\{\exp(-tW^T BA^{-1} BW/\sigma^2)\}\,dt$$

$$= \int_0^\infty E\left\{\exp\left(-t \sum_{j=1}^P M_j^2 d_j^2/\sigma^2\right)\right\}\,dt \qquad (4.10)$$

where $M = SW \sim N(\Lambda, \sigma^2 D^{-1})$, $\Lambda = S\gamma$ and S is chosen so that $SB^{-1}S^T = D^{-1} = \mathrm{diag}(d_1^{-1}, \ldots, d_P^{-1})$ and $SA^{-1}S^T = I_p$. Further, $M_j^2 d_j/\sigma^2$, $j = 1, \ldots, p$, are independently distributed as $\chi_1^2(\Lambda_j^2 d_j/(2\sigma^2))$. Hence,

$$E\left\{\exp\left(-t \sum_{j=1}^P M_j^2 d_j^2/\sigma^2\right)\right\}$$

$$= \prod_{j=1}^P E\{\exp(-tM_j^2 d_j^2/\sigma^2)\}$$

$$= \prod_{j=1}^P (1 + 2td_j)^{-1/2} \exp\left\{-\sigma^{-2}t \sum_{k=1}^P \frac{\Lambda_k^2 d_k^2}{1 + 2td_k}\right\}$$

$$= |I_p + 2tA^{-1}B|^{-1/2} \exp\{-\sigma^{-2}t\gamma^T B(A + 2tB)^{-1} B\gamma\} \qquad (4.11)$$

The result in (iii) follows from (4.7) and (4.9)–(4.11).

(iv) For the case in which $\beta \neq \lambda$, fixed, by (4.8),

$$n^2 E\{\|\hat{\beta}_n(\phi) - \lambda\|^{-2}\}$$

$$= \sigma^{-2} \int_0^\infty \left|I_p + 2\frac{t}{n} Q_n^{-1} n^{-1}(X^T X)\right|^{-1/2} \exp\left\{-\sigma^{-2}t(\beta - \lambda)^T n^{-1}(X^T X)\right.$$

$$\times \left(Q_n + 2\frac{t}{n}n^{-1}(X^TX))^{-1}n^{-1}(X^TX)(\beta - \lambda) \right\} dt$$

$$\to \sigma^{-2} \int_0^\infty \exp\{ - \sigma^{-2}t(\beta - \lambda)^T BA^{-1}B(\beta - \lambda)\} dt$$

$$= \{(\beta - \lambda)^T BA^{-1}B(\beta - \lambda)\}^{-1}. \tag{4.12}$$

The result in (iv) then follows from (4.7) and (4.12).

From Theorem 4.1, for any $0 < b < 2(p - 2)$, the risk of $\delta_n^s(\phi)$ is smaller than that of $\hat{\beta}_n(\phi)$ up to the order n^{-1}, at all values of β in the neighborhood of λ. Here, the risk difference is maximized at $b = p - 2$. However, when $\beta \neq \lambda$, fixed, the dominance of $\delta_n^s(\phi)$ over $\hat{\beta}_n(\phi)$ is only of the order n^{-2}. Here, also, the risk difference is maximized at $b = p - 2$. Consequently, the risk dominance persists even for large samples, which, given Theorem 3.1, is not surprising.

The risk expansion derived in Theorem 4.1 gives the order of dominance of $\delta_n^s(\phi)$ over $\hat{\beta}_n(\phi)$ for large samples. We can now show that a similar asymptotic dominance of δ_n^s over $\hat{\beta}_n$ obtains when ϕ is unknown.

4.2 Autoregressive Parameter Unknown

If ϕ is unknown we need only replace ϕ in (1.10) and (2.4) by $\hat{\phi}_n$ given in (1.13) and $\Sigma_n = X^TX - X^TY_l[Y_l^TY_l]^{-1}Y_l^TX$ in (2.4) and (2.5). Denote the resulting estimates $\hat{\beta}_n(\hat{\phi}_n)$ and $\delta_n^s(\hat{\phi}_n)$ by $\hat{\beta}_n$ and δ_n^s, respectively. In this section we establish the asymptotic risk dominance of δ_n^s over $\hat{\beta}_n$ in a sense to be stated later. The limit distribution of $\hat{\beta}_n$ can be found in, for example Chapter 9, Section 8 of Fuller (1976), and is stated again in Theorem 4.2. We shall also require the limit distribution of δ_n^s which is derived below in Theorem 4.3.

Theorem 4.2. Assume that in addition to (4.1) and (4.2) the following conditions are satisfied:

$$n^{-1}\sum_{t=1}^n X_t y_{t-1} = n^{-1}X^TY_l \to l \tag{4.13}$$

and

$$n^{-1}\sum_{t=1}^n y_{t-1}^2 = n^{-1}Y_l^TY_l \to m, \tag{4.14}$$

where both convergences are in probability, as $n \to \infty$. We then have

$$\text{(i)} \quad \sqrt{n} \begin{bmatrix} \hat{\beta}_n - \beta \\ \hat{\phi}_n - \phi \end{bmatrix} \to N\left(\begin{bmatrix} 0 \\ 0 \end{bmatrix}, \sigma^2 \begin{bmatrix} B & l \\ l^T & m \end{bmatrix}^{-1} \right)$$

in distribution, as $n \to \infty$, and

$$\text{(ii)} \quad nL_n(\hat{\beta}_n, \beta) \to U^T A U$$

in distribution, as $n \to \infty$, where L_n is the loss function defined in (2.1), $U \sim N(0, \sigma^2 \Sigma^{-1})$ and

$$\Sigma^{-1} = B^{-1} + B^{-1}l(m - l^T B^{-1}l)^{-1}l^T B^{-1} \qquad (4.15)$$

$$= (B - lm^{-1}l^T)^{-1}$$

Proof: The result in (i) is an immediate consequence of (4.1), (4.2), (4.13) and (4.14) (See, for example, Anderson (1971), chapter 5). To prove (ii) observe that

$$\begin{bmatrix} B & l \\ l^T & m \end{bmatrix}^{-1} = \begin{bmatrix} \Sigma^{-1} & -B^{-1}l(m - l^T B^{-1}l)^{-1} \\ -(m - l^T B^{-1}l)^{-1}l^T B^{-1} & (m - l^T B^{-1}l)^{-1} \end{bmatrix}$$
$$(4.16)$$

Consequently, from (4.16) and (i),

$$\sqrt{n}\,(\hat{\beta}_n - \beta) \to N(0, \sigma^2 \Sigma^{-1}) \qquad (4.17)$$

in distribution, as $n \to \infty$. Hence as $n \to \infty$

$$nL_n(\hat{\beta}_n, \beta) = n(\hat{\beta}_n - \beta)^T Q_n(\hat{\beta}_n - \beta)$$

$$= [\sqrt{n}(\hat{\beta}_n - \beta)]^T Q_n [\sqrt{n}\,(\hat{\beta}_n - \beta)]$$

$$\to U^T A U$$

in distribution.

The next theorem establishes the asymptotic distribution of the shrinkage estimate δ_n^s.

Theorem 4.3. Under the conditions (4.1), (4.2), (4.13) and (4.14) we have an $n \to \infty$, in distribution,

$$\sqrt{n}(\delta_n^s - \beta) \to \begin{cases} U, & \text{for } \beta \neq \lambda \\ g(U) & \text{for } \beta = \lambda \end{cases} \qquad (4.18)$$

where U is defined as in Theorem 4.2, and $g(u)$ is defined as

$$g(u) = \left[I_p - \frac{b\sigma^2}{u^T \Sigma A^{-1} \Sigma u} A^{-1} \Sigma \right] u. \tag{4.19}$$

Proof: Note that

$$\sqrt{n}(\delta_n^s - \beta) = \sqrt{n}(\hat{\beta}_n - \beta) - C_n \sqrt{n}(\hat{\beta}_n - \lambda) \tag{4.20}$$

where

$$C_n = C_n(\hat{\phi}) = b\hat{\sigma}_n^2 \|\hat{\beta}_n - \lambda\|^{-2} Q_n^{-1} \Sigma_n \tag{4.21}$$

with $\hat{\sigma}_n^2 = \hat{\sigma}_n^2(\hat{\phi}_n)$ and $\Sigma_n = X^T X - X^T Y_l [Y_l^T Y_l]^{-1} Y_l^T X$.
   At this point, let us note that, via standard argument,

$$\hat{\sigma}_n^2 \to \sigma^2 \tag{4.22}$$

in probability, as $n \to \infty$. This result will be of use in proving (4.18).
   Now, when $\beta = \lambda$,

$$C_n \sqrt{n}(\hat{\beta}_n - \lambda) = b \frac{\hat{\sigma}_n^2}{n^{-1}\|\hat{\beta}_n - \beta\|^2} Q_n^{-1}[n^{-1}\Sigma_n]\sqrt{n}(\hat{\beta}_n - \beta) \tag{4.23}$$

with

$$n^{-1}\|\hat{\beta}_n - \beta\|^2 = \sqrt{n}(\hat{\beta}_n - \beta)^T [n^{-1}\Sigma_n]Q_n^{-1}[n^{-1}\Sigma_n]\sqrt{n}(\hat{\beta}_n - \beta) \tag{4.24}$$

Consequently, applying (4.1), (4.2), (4.17), and (4.22)–(4.24) to (4.20) and combining it with (4.25) gives us

$$\sqrt{n}(\delta_n^s - \beta) \to g(U) \tag{4.25}$$

in distribution, as $n \to \infty$.
   Next, consider the case $\beta \neq \lambda$. Then

$$C_n \sqrt{n}(\hat{\beta}_n - \lambda) = n^{-1/2} b \frac{\hat{\sigma}_n^2}{n^{-2}\|\hat{\beta}_n - \lambda\|^2} Q_n^{-1}[n^{-1}\Sigma_n](\hat{\beta}_n - \lambda) \tag{4.26}$$

with

$$n^{-2}\|\hat{\beta}_n - \lambda\|^2 = (\hat{\beta}_n - \lambda)^T [n^{-1}\Sigma_n]Q_n^{-1}[n^{-1}\Sigma_n(\hat{\beta}_n - \lambda)] \tag{4.27}$$

Consequently, using the fact that

$$\hat{\beta}_n \to \beta \tag{4.28}$$

in probability, as $n \to \infty$, in conjunction with (4.1), (4.2), (4.22) and (4.26)–(4.28) we have

$$C_n \sqrt{n}(\hat{\beta}_n - \lambda) \to 0 \qquad (4.29)$$

in probability, as $n \to \infty$. Using (4.29) with (4.17) and (4.20) we obtain

$$\sqrt{n}(\delta_n^s - \beta) \to U \qquad (4.30)$$

in distribution, as $n \to \infty$. Results (4.25) and (4.30) give us (4.18).

Next, for the case of unknown ϕ, we shall use a weaker concept of asymptotic risk rather than the limiting risk to compare the shrinkage estimator δ_n^s to the maximum likelihood estimate $\hat{\beta}_n$ for large samples. This is akin to the usual criterion of using the variance of the limit distribution rather than the limit of the variance of an estimator for first order asymptotic comparisons of estimators. For this purpose we shall use the limit distributions derived in Theorems 4.2 and 4.3. For the case of known ϕ we have already obtained in Theorem 4.1 the relative efficiency of $\delta_n^s(\phi)$ with respect to $\hat{\beta}_n(\phi)$ using the concept of asymptotic risk rather than the limiting risk.

We have seen that, in Theorem 4.2,

$$nL_n(\hat{\beta}_n, \beta) \to U^T A U$$

in distribution, as $n \to \infty$. Consequently, the asymptotic risk of $\hat{\beta}_n$ will be defined as

$$AR(\hat{\beta}_n, \beta) = E(U^T A U) = \sigma^2 tr(A\Sigma^{-1}). \qquad (4.31)$$

Next, consider

$$nL_n(\delta_n^s, \beta) = [\sqrt{n}(\delta_n^s - \beta)]^T Q_n [\sqrt{n}(\delta_n^s - \beta)] \qquad (4.32)$$

Therefore, when $\beta \neq \lambda$, from Theorem 4.3,

$$nL_n(\delta_n^s, \beta) \to U^T A U \qquad (4.33)$$

in distribution, as $n \to \infty$. Consequently, for $\beta \neq \lambda$, from (4.33),

$$AR(\delta_n^s, \beta) = \sigma^2 tr(A\Sigma^{-1}) = AR(\hat{\beta}_n, \beta) \qquad (4.34)$$

However, when $\beta = \lambda$, from (4.32) and Theorem 4.3,

$$nL_n(\delta_n^s, \beta) \to [g(U)]^T A[g(U)] \qquad (4.35)$$

in distribution, as $n \to \infty$. Here,

$$[g(U)]^T A[g(U)] = U^T A U + b^2 \frac{\sigma^4}{U^T \Sigma A^{-1} \Sigma U} - 2b\sigma^2 \frac{U^T \Sigma U}{U^T \Sigma A^{-1} \Sigma U} \qquad (4.36)$$

Again, by the simultaneous diagonalization theorem there exists a positive definite matrix R such that $R\Sigma^{-1}R^T = I_p$ and $RA^{-1}R^T = D^{-1} = \text{diag}(d_1^{-1}, \ldots, d_p^{-1})$. Then, setting $M = RU \sim N(0, \sigma^2 I_p)$, we have

$$E\left[\frac{U^T\Sigma U}{U^T\Sigma A^{-1}\Sigma U}\right] = E\left[\frac{(RU)^T(R^T)^{-1}\Sigma R^{-1}(RU)}{(RU)^T(R^T)^{-1}\Sigma R^{-1}RA^{-1}R^T(R^T)^{-1}\Sigma R^{-1}(RU)}\right]$$

$$= E\left[\frac{M^TM}{M^TD^{-1}M}\right] = \sum_{j=1}^{p} E\left[\frac{M_j^2}{\Sigma_{k=1}^{p} M_k^2/d_k}\right] \qquad (4.37)$$

Now, for each $j = 1, \ldots, p$, by Stein's identity,

$$E\left[\frac{M_j^2}{\Sigma_{k=1}^{p} M_k^2/d_k}\right] = \sigma^2 E\left[\frac{M_j}{\Sigma_{i=1}^{p} M_k^2/d_k}\frac{M_j}{\sigma^2}\right]$$

$$= \sigma^2 E\left[\frac{\partial}{\partial M_j}\frac{M_j}{\Sigma_{k=1}^{p} M_k^2/d_k}\right]$$

$$= \sigma^2 E\left[\frac{1}{\Sigma_{k=1}^{p} M_k^2/d_k} - 2\frac{M_j^2/d_j}{[\Sigma_{k=1}^{p} M_k^2/d_k]^2}\right]$$

$$= \sigma^2 E\left[\frac{1}{M^TD^{-1}M} - 2\frac{M_j^2/d_j}{[M^TD^{-1}M]^2}\right]. \qquad (4.38)$$

Hence, by (4.37) and (4.38),

$$E\left[\frac{U^T\Sigma U}{U^T\Sigma A^{-1}\Sigma U}\right] = (p-2)\sigma^2 E\left[\frac{1}{U^T\Sigma A^{-1}\Sigma U}\right] \qquad (4.39)$$

Also, following arguments similar to those found in (4.8)

$$E[(U^T\Sigma A^{-1}\Sigma U)^{-1}] = E\left[\left(\sum_{k=1}^{p} M_k^2/d_k\right)^{-1}\right]$$

$$= \sigma^{-2}\int_0^{\infty} E\left\{\exp\left(-t\sum_{k=1}^{p}\frac{M_k^2}{\sigma^2}d_k^{-1}\right)\right\} dt = \sigma^{-2}\int_0^{\infty}\prod_{k=1}^{p} E\left\{\exp\left(-t\frac{M_k^2}{\sigma^2}d_k^{-1}\right)\right\} dt$$

$$= \sigma^{-2}\int_0^{\infty}|I_p + 2tD^{-1}|^{-1/2} dt = \sigma^{-2}\int_0^{\infty}|I_p + 2tA^{-1}\Sigma|^{-1/2}dt. \qquad (4.40)$$

Consequently, by (4.35), (4.36), (4.39) and (4.40), for $\beta = \lambda$,

$$AR(\delta_n^s, \beta) = \sigma^2 \operatorname{tr}(A\Sigma^{-1}) - \sigma^2 b(2(p-2) - b) \int_0^\infty |I_p + 2tA^{-1}\Sigma|^{-1/2} \, dt$$

$$(4.41)$$

If we compare (4.31), (4.34) and (4.41) we find that

$$AR(\delta_n^s, \beta) \begin{cases} = AR(\hat{\beta}_n, \beta), & \text{for } \beta \neq \lambda \\ < AR(\hat{\beta}_n, \beta), & \text{for } \beta = \lambda, 0 < b < 2(p-2). \end{cases} \quad (4.42)$$

Further, it can be shown that, for the case $\beta = \lambda + n^{-1/2}\gamma$, we have

$$nL_n(\delta_n^s, \beta) \to U^T A U + b^2 \frac{\sigma^4}{(U+\gamma)^T \Sigma A^{-1} \Sigma (U+\gamma)}$$

$$- 2b\sigma^2 \frac{(U+\gamma)^T \Sigma U}{(U+\gamma)^T \Sigma A^{-1} \Sigma (U+\gamma)} \quad (4.43)$$

in distribution, as $n \to \infty$. We omit the details. Consequently, for $\beta = \lambda + n^{-1/2}\gamma$,

$$AR(\delta_n^s, \beta) = \sigma^4 b\{2(p-2) - b\} E\{(\Gamma^T \Sigma A^{-1} \Sigma \Gamma)^{-1}\} + \sigma^2 \operatorname{tr}(A\Sigma^{-1}) \quad (4.44)$$

where $\Gamma \sim N(\gamma, \sigma^2 \Sigma^{-1})$. From (4.44) it is clear that $AR(\delta_n^s, \beta)$ is less than $AR(\hat{\beta}_n, \beta)$ for $0 < b < 2(p-2)$, when $\beta = \lambda + n^{-1/2}\gamma$.

Plainly stated, the asymptotic risk of δ_n^s is less than that of $\hat{\beta}_n$ at, and in the neighborhood of $\beta = \lambda$. On the other hand, both estimators have the same asymptotic risk for $\beta \neq \lambda$, fixed. These results regarding the comparison of asymptotic risks of δ_n^s and $\hat{\beta}_n$ for the case when ϕ is unknown are compatible with the stronger results on the comparison of limiting risks given in Theorem 4.1 where ϕ is assumed known. It may be further noted that the convergence of δ_n^s is nonuniform at and around the value $\beta = \lambda$. This is a typical undesirable asymptotic feature shared by most shrinkage estimators.

ACKNOWLEDGMENT

I. V. Basawa's work was partially supported by a grant from the Office of Naval Research.

REFERENCES

Anderson, T. W. (1971). *Statistical Analysis of Time Series*. Wiley, New York.

Baranchik, A. J. (1970). A family of minimax estimators of the mean of a multivariate normal distribution. *Ann. Math. Statist. 41*, 642–645.

Berger, J. O. (1976). Admissible minimax estimation of a multivariate normal mean with arbitrary quadratic loss, *The Annals of Statistics*, 4, 223–226.

Berger, J. O. and Bock, M. E. (1976a). Combining independent normal mean estimation problems with unknown variances, *The Annals of Statistics*, 4, 642–648.

Berger, J. O. and Bock, M. E. (1976b). Improved minimax estimators of normal mean vectors for certain types of covariance matrices. Proceedings of the 2nd Conference on Statistical Decision Theory and Related Topics, 19–36. Academic Press, New York.

Berger, J. Bock, M. E., Brown, L. D., Casella, G., and Gleser, L. (1977). Minimax estimation of a normal mean vector for arbitrary quadratic loss and unknown covariance matrix', *The Annals of Statistics*, 5, 763–771.

Berger, J. and Haff, L. R. (1983). A class of minimax estimators of a normal mean vector for arbitrary quadratic loss and unknown covariance matrix, *Statistics and Decisions*, 1, 105–129.

Efron, B. and Morris, C. (1976). Families of minimax estimators of the mean of a multivariate normal distribution, *The Annals of Statistics*, 4, 11–21.

Fuller, W. A. (1976). *Introduction to Statistical Time Series*. Wiley, New York.

James, W. and Stein, C. (1961). Estimation with quadratic loss. *Proc. 4th Berkeley Symp. Math. Statist. Probab. 1*, 361–379. Univ. of Calif. Press.

Rao, C. R. (1973). *Linear Statistical Inference*. Wiley, New York.

Sclove, S. L. (1968). Improved estimators for coefficients in linear regression. *J. Amer. Statist. Assoc., 63*, 596–606.

Strawderman, W. E. (1971). Proper Bayes minimax estimators of the multivariate normal mean. *Ann. Math. Statist., 42*, 385–388.

11
Maximum Probability Estimation for an Autoregressive Process

Lionel Weiss
School of Operations Research and Industrial Engineering, Cornell University, Ithaca, New York

We observe $X_1, \ldots, X_n$, where $X_i = \theta_1 X_{i-1} + Y_i$, where X_0 is defined as zero, and $Y_1, \ldots, Y_n$ are unobservable random variables, independent, each normal with mean zero and variance θ_2; θ_1 and θ_2 are both unknown and are to be estimated. By using maximum probability estimators, it is shown that the maximum likelihood estimators of the parameters have certain optimal properties.

1. INTRODUCTION

Maximum probability estimators were developed in Weiss and Wolfowitz (1967, 1968, 1970, 1974). They represented an attempt to explain why maximum likelihood estimators work so well in so many cases. The theory was predominantly asymptotic. A brief description of this asymptotic theory follows. There is an index n, which takes positive integral values, and we examine what happens as n increases. For each n, $X(n)$ is the random vector to be observed. The joint density function for the components of $X(n)$ is $f_n(x(n); \theta_1, \ldots, \theta_m)$, where $\theta_1, \ldots, \theta_m$ are unknown parameters to be estimated. For each n and each vector

$(\theta_1, \ldots, \theta_m)$, $R_n(\theta_1, \ldots, \theta_m)$ is a bounded measurable subset of m-dimensional space, containing the m-dimensional origin $(0, \ldots, 0)$, and as n increases $R_n(\theta_1, \ldots, \theta_m)$ shrinks toward this origin. For any given values $D_1, \ldots, D_m$, let $R_n^*(D_1, \ldots, D_m)$ denote the set of vectors $(\theta_1, \ldots, \theta_m)$ such that the vector $(D_1 - \theta_1, \ldots, D_m - \theta_m)$ is in $R_n(\theta_1, \ldots, \theta_m)$. Let $\tilde{\theta}_1(n), \ldots, \tilde{\theta}_m(n)$ denote the values of $D_1, \ldots, D_m$ which maximize the integral

$$\int \cdots \int_{R_n^*(D_1, \ldots, D_m)} f_n(X(n); \theta_1, \ldots, \theta_m) \, d\theta_1 \ldots d\theta_m$$

$\tilde{\theta}_1(n), \ldots, \tilde{\theta}_m(n)$ are called "maximum probability estimators with respect to $R_n(\theta_1, \ldots, \theta_m)$" because if $\theta_1, \ldots, \theta_m$ can be estimated with increasing accuracy as n increases, then among all estimators satisfying a reasonable regularity condition, $(\tilde{\theta}_1(n), \ldots, \tilde{\theta}_m(n))$ gives the highest asymptotic probability that the vector (estimator of $\theta_1 - \theta_1, \ldots$, estimator of $\theta_m - \theta_m$) falls in $R_n(\theta_1, \ldots, \theta_m)$.

In Weiss (1983) small-sample properties of maximum probability estimators were investigated. Now, we apply the theory of maximum probability estimators to the following estimation problem. $Y_1, \ldots, Y_n$ are independent and identically distributed, each normal with mean zero and unknown positive variance θ_2. We do not observe $Y_1, \ldots, Y_n$. What we observe are $X_1, \ldots, X_n$, where $X_i = \theta_1 X_{i-1} + Y_i$ with X_0 defined as zero. θ_1 is an unknown parameter. The problem is to estimate θ_1 and θ_2.

The joint density function for $X_1, \ldots, X_n$ is $(2\pi\theta_2)^{-n/2}$ $\exp[(-1/2\theta_2) \sum_{i=1}^n (x_i - \theta_1 x_{i-1})^2]$ and the maximum likelihood estimators are

$$\hat{\theta}_1 = \frac{\sum_{i=1}^n X_i X_{i-1}}{\sum_{i=1}^n X_{i-1}^2}$$

$$\hat{\theta}_2 = \frac{1}{n} \sum_{i=1}^n (X_i - \hat{\theta}_1 X_{i-1})^2$$

In Weiss (1983) and Weiss and Wolfowitz (1968) this process was studied for the case where θ_2 was known to be equal to one, and so only θ_1 had to be estimated.

2 THE CASE $|\theta_1| < 1$

If $|\theta_1| < 1$, it is easily verified that the conditions of Weiss (1973) are satisfied, and therefore the maximum likelihood estimators have the following properties: $n^{1/2}(\hat{\theta}_1 - \theta_1)$ and $n^{1/2}(\hat{\theta}_2 - \theta_2)$ have asymptotically a bivariate normal distribution, with zero means, respective variances $1 - \theta_1^2$, $2\theta_2^2$, and covariance zero. Furthermore, $\hat{\theta}_1$, $\hat{\theta}_2$ have the same asymptotic properties as maximum probability estimators with respect to any convex region symmetric about the origin $(0, 0)$.

3 THE ESTIMATION OF θ_1

Knowing θ_2 is of no advantage in estimating θ_1. That is, a statistician who does not know θ_2 can estimate θ_1 as well as a statistician who knows the value of θ_2. To see this, imagine that unknown to us, each Y_i is multiplied by a nonzero value c. This replaces θ_2 by $c^2\theta_2$, and multiplies each X_i by c, but does not change either θ_1 or $\hat{\theta}_1$. The fact that knowing θ_2 is of no relevance for estimating θ_1 explains the form of the asymptotic covariance matrix in Section 2.

In Weiss (1983) the problem of estimating θ_1 when θ_2 is known to be one and θ_1 is completely unknown was studied. In this case, the maximum likelihood estimator $\hat{\theta}_1$ is also the maximum probability estimator with respect to any region $R_n(\theta_1)$ symmetric around zero, and it was shown that $\hat{\theta}_1$ is uniformly arbitrarily close to a Bayes estimator, for each $n > 2$. From the discussion above, it follows that these desirable properties of $\hat{\theta}_1$ also hold if θ_2 is unknown.

4 ASYMPTOTIC PROPERTIES OF $\hat{\theta}_2$ WHEN $|\theta_1| \geq 1$

In this section, we show that if $|\theta_1| \geq 1$, the asymptotic distribution of $\hat{\theta}_2$ is the same as that of the maximum likelihood estimator of θ_2 which would be available to somebody who knew the value of θ_1. Denote by θ_2^* the maximum likelihood estimator of θ_2 when θ_1 is known. $\theta_2^* = 1/n \sum_{i=1}^{n} (X_i - \theta_1 X_{i-1})^2$. Since $n\theta_2^*/\theta_2$ has a chi-square distribution with n degrees of freedom, it follows that $(n/2\theta_2^2)^{1/2} (\theta_2^* - \theta_2)$ is asymptotically standard normal.

Let R denote $\sum_{i=1}^{n} X_{i-1}Y_i$ and let S denote $\sum_{i=1}^{n} X_{i-1}^2$. The following relations are easily verified:

$$\hat{\theta}_1 = \theta_1 + \frac{R}{S} \tag{4.1}$$

$$\hat{\theta}_2 - \theta_2^* = \frac{-(\hat{\theta}_1 - \theta_1)^2 S}{n} \tag{4.2}$$

We will show that if $|\theta_1| \geqslant 1$, then $(\hat{\theta}_1 - \theta_1)^2 S/n^{1/2}$ converges to zero in probability as n increases. Considering (4.2) and the asymptotic distribution of θ_2^*, this will prove that the asymptotic distribution of $(n/2\theta_2^2)^{1/2}(\hat{\theta}_2 - \theta_2)$ is also standard normal.

First we discuss the case where $|\theta_1| > 1$. Since $X_i = \Sigma_{j=1}^i \theta_1^{i-j} Y_i$ for $i \geqslant 1$, X_i has a normal distribution with mean zero and variance $\theta_2[(1 - \theta_1^{2i})/(1 - \theta_1^2)]$. Since X_{i-1} is independent of Y_i, it follows easily that R has mean zero and variance $[\theta_2^2/(1 - \theta_1^2)][n - 1 - (\theta_1^2 - \theta_1^{2n})/(1 - \theta_1^2)]$. It is easily shown that $S = (1 - \theta_1^2)^{-1}[-X_n^2 + \Sigma_{i=1}^n Y_i^2 + 2\theta_1 R]$. Thus R is $O_p(|\theta_1|^n)$, X_n is $O_p(|\theta_1|^{2n})$, and $\Sigma_{i=1}^n Y_i^2$ is $O_p(n)$. It follows that S is $O_p(|\theta_1|^{2n})$. From (4.1) we then find that $\hat{\theta}_1 - \theta_1$ is $O_p(|\theta_1|^{-n})$. Then $(\hat{\theta}_1 - \theta_1)^2 S/n^{1/2}$ converges to zero in probability as n increases.

Now we look at the case where $|\theta_1| = 1$. We carry out the details for the case $\theta_1 = 1$, the case $\theta_1 = -1$ being essentially the same. In this case R has mean zero and variance $\theta_2^2 n(n - 1)/2$. S has mean $\theta_2 n(n - 1)/2$. In computing the variance of S, we use the fact that if $j > i$, then $X_j = X_i + (Y_{i+1} + \cdots + Y_j)$, and $(Y_{i+1} + \cdots + Y_j)$ is independent of X_i. A straightforward calculation then shows that the variance of S is $\theta_2^2 n^4/12 + \theta_2^2 O(n^3)$. We can write R as $\theta_2[n(n - 1)/2]^{1/2} R'$ where R' is $O_p(1)$, and we can write S as $\theta_2 n(n - 1)/2 + \theta_2 n^2 S'/(12)^{1/2}$ where S' is $O_p(1)$. It follows directly that $\hat{\theta}_1 - \theta_1 = R/S$ is $O_p(1/n)$, and then $(\hat{\theta}_1 - \theta_1)^2 S/n^{1/2} = O_p(n^{-1/2})$. This completes the demonstration.

REFERENCES

L. Weiss (1973). Asymptotic properties of maximum likelihood estimators in some nonstandard cases, Part II, *J. Amer. Statist. Assoc.*, **68**, 728–730.

L. Weiss (1983). Small-sample properties of maximum probability estimators, *Stochastic Processes and their Applications*, **14**, 267–277.

L. Weiss and J. Wolfowitz (1967). Maximum probability estimators, *Ann. Inst. Stat. Math.* **19**, 193–206.

L. Weiss and J. Wolfowitz, (1968). Generalized maximum likelihood estimators, *Teoriya Vyeroyatnostey*, **13**, 68–97.

L. Weiss and J. Wolfowitz (1970). Maximum probability estimators and asymptotic sufficiency, *Ann. Inst. Stat. Math.*, **22**, 225–244.

L. Weiss and J. Wolfowitz (1974). Maximum Probability Estimators and Related Topics. Springer, Berlin.

Index

For Product Safety Concerns and Information please contact our EU
representative GPSR@taylorandfrancis.com
Taylor & Francis Verlag GmbH, Kaufingerstraße 24, 80331 München, Germany

www.ingramcontent.com/pod-product-compliance
Ingram Content Group UK Ltd.
Pitfield, Milton Keynes, MK11 3LW, UK
UKHW040927180425
457613UK00004B/45